T0397705

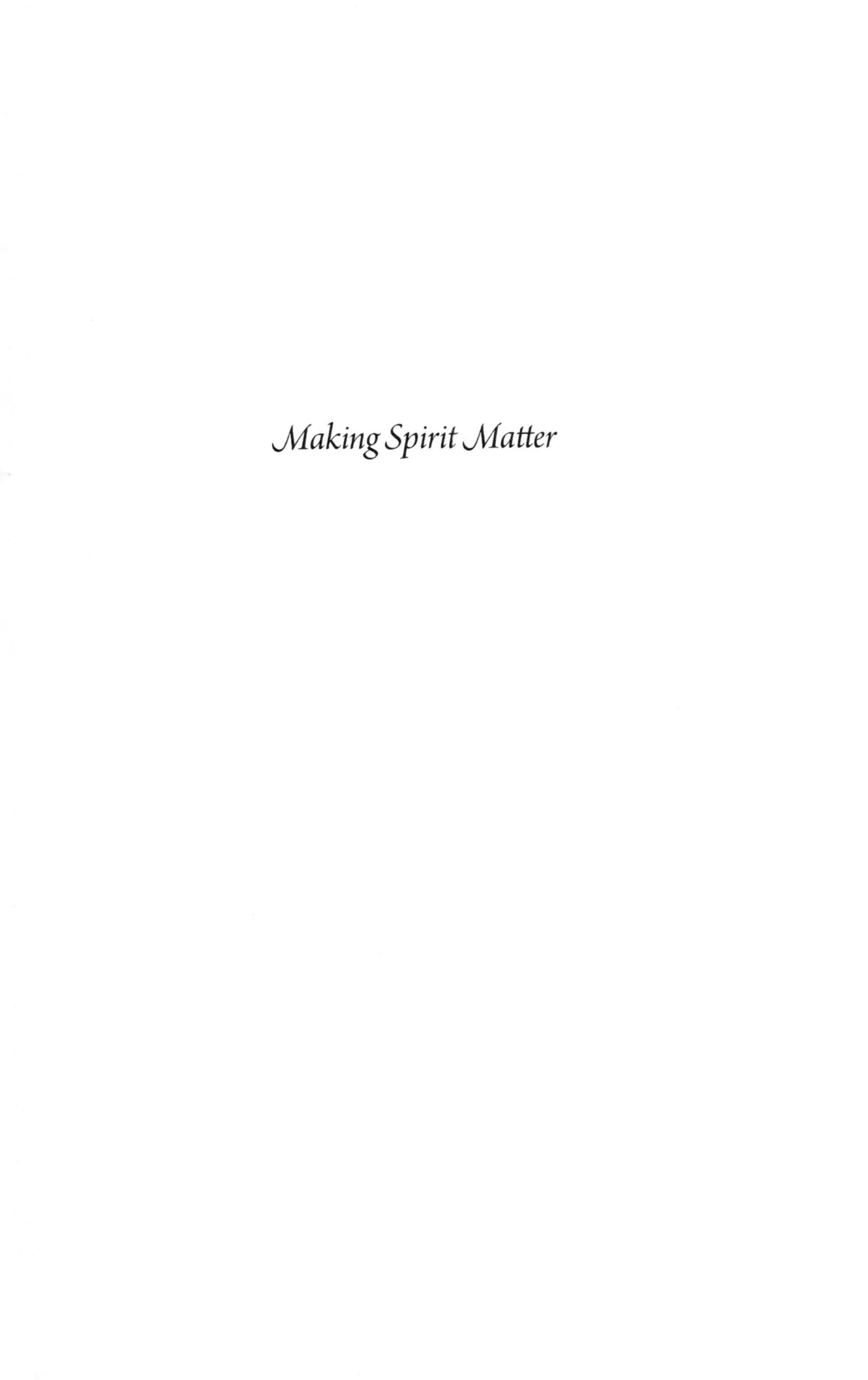

Making Spirit Matter

Making Spirit Matter

NEUROLOGY, PSYCHOLOGY, AND SELFHOOD IN MODERN FRANCE

Larry Sommer McGrath

The University of Chicago Press CHICAGO AND LONDON

PUBLICATION OF THIS BOOK HAS BEEN AIDED
BY A GRANT FROM THE BEVINGTON FUND.

The University of Chicago Press, Chicago 60637
The University of Chicago Press, Ltd., London

Published 2020
Printed in the United States of America

29 28 27 26 25 24 23 22 21 20   1 2 3 4 5

ISBN-13: 978-0-226-69979-0 (cloth)
ISBN-13: 978-0-226-69982-0 (paper)
ISBN-13: 978-0-226-69996-7 (e-book)
DOI: https://doi.org/10.7208/chicago/9780226699967.001.0001

Library of Congress Cataloging-in-Publication Data

Names: McGrath, Larry Sommer, author.
Title: Making spirit matter : neurology, psychology, and selfhood in modern France / Larry Sommer McGrath.
Description: Chicago ; London : The University of Chicago Press, 2020. | Includes bibliographical references and index.
Identifiers: LCCN 2020001642 | ISBN 9780226699790 (cloth) | ISBN 9780226699820 (paperback) | ISBN 9780226699967 (ebook)
Subjects: LCSH: Spiritualism (Philosophy) | Mind and body—France. | Philosophy, French—19th century. | Philosophy, French—20th century. | Philosophy and science—France. | France—Intellectual life—19th century. | France—Intellectual life—20th century.
Classification: LCC B841 .M36 2020 | DDC 128/.2094409034—dc23
LC record available at https://lccn.loc.gov/2020001642

♾ This paper meets the requirements of ANSI/NISO Z39.48-1992 (Permanence of Paper).

Contents

Introduction

How do our thoughts interact with our bodies? Do we have freedom over and above the brain's operations? Or do we live in thrall to our biology? These problems achieved remarkable urgency in France during the nineteenth century, a period when the brain came to be taken seriously as a special organ. For many scientists, it was not only the locus of thinking but also the linchpin of a materialist worldview that excised consciousness and free will from the natural world. God, so it followed, was dispelled as an illusion as well. The workings of the nervous system ceased to be shrouded in mystery thanks to the emergent fields of psychology and neurology, which unveiled the complex network facilitating the body's sensory and motor connections. Those workings—electrical, energetic, and material—stirred widespread excitement and anxiety in the prospect that humans' spiritual powers—including reason, reflection, and action—were made up solely of physical matter. Today, the neurosciences shoulder a similar aspiration to reveal the material underpinnings of subjectivity. Yet, the history of the mind-body problem is so fascinating in modern France because that was not entirely the case. The early brain sciences were entangled with the immaterial. Incredibly, the deeper that French neurologists and psychologists explored the functions of the nervous system, the more they confronted the stubborn problem of spirit.

The spiritual and the material have been interwoven in French history, the subject of this book. Although often framed in opposition, these braided concepts are indispensable to how French people think of themselves. *L'esprit* and *la matière* are the basic terms with which one makes oneself intelligible—that is, gives account of what "I am." This "I" inhabits a body composed of matter. It takes up space and resists our touch. What it means for someone also to have a spirit, however, poses difficulties. To be clear, the term does not refer to ghosts. Although spirit often has a mystical connotation, it does not for my purposes in this book. The French word *l'esprit* signifies our inner realm of experience. Etymologically, that realm is immaterial, extending back to the Latin *spiritus,* meaning "breath," as well as the Greek πνεῦμα (pneuma): the air sustaining the body. In our ordinary language, the spirit of the law refers to the intent conveyed but not legible in the letter. Although today spirit features prominently among enthusiasts of yoga, Tantra, and other practices originating in Eastern cultures, such associations would not be the first to come to mind in France. Some come close: "cognition," "consciousness," or "mind." Nevertheless, none of these concepts was nearly so integral as spirit was to understandings of selfhood across science and society over the past two hundred years.

It might seem striking that spirit—a concept pregnant with ethereal meaning—played a formative part in the nascent brain sciences. But that's precisely what I aim to show in this book. Far from having sealed the fate of metaphysics and theology, insights from neurology and psychology instead made it possible to envision our spiritual powers afresh. By the beginning of the twentieth century, French cultural life was consumed with reports from laboratory experiments on the nervous system and from clinical observations of people with brain disorders. Brain autopsies revealed the nervous tissues governing language, speech, and memory. Maps of the organ's regions captivated philosophers and religious thinkers, medical doctors and mathematicians, social reformers and policy makers. What diverse readers in France found so captivating about brain literature was how frequently it confounded the reductive ambition to locate the hub of mental operations in the space between our ears. As people learned more about the pathways of electrical signals traversing our body, connecting the brain and spinal cord to the muscles and skin, unanswered questions overflowed the bounds of science. Novel possibilities emerged. Specifically, the nervous system offered an empirical point of departure to imagine the corporeal points of contact where spirit converged with matter. The constellation of figures in this book did just that by crafting inventive accounts of freedom, thought, action, and revelation, all of which transformed how people understood the body's creative role in human experience.

Spirit's renaissance came about thanks to monumental upheavals in French society. During the second half of the nineteenth century, the government invested in scientific advancement in order to rejuvenate the social order. In the wake of a humiliating defeat in the Franco-Prussian War of 1870, the French Third Republic took shape. It was a delicate experiment that proved to be the country's longest democratic period (lasting seventy years until the Nazi occupation of 1940). In their endeavors to construct the nation anew, social reformers turned to science as a key to surpassing the technological prowess of European foes—chiefly, the newly unified German state. Scientific laboratories were built in the universities; a mass readership took shape around new scientific journals; and the French education system made scientific instruction a priority.[1] Spirit was allied with science. And together they became a focal point of the official school curriculum.

In an era before television, radio, and the internet — modern conduits of ideology — education functioned as the primary ideological mouthpiece of the state. Hardly neutral, state-sponsored education was pressed into the service of molding the ideal "rational" French citizen whose fidelity to the *patrie* would buttress the country's fragile democracy. In the final decades of the nineteenth century, modernizing reforms flowed from parliament, extending mandatory schooling across the nation to standardize the French language, to train future workers, and — crucially — to inculcate a vision of selfhood premised on neurophysiology. Schools enforced a mixed ideology committed both to future scientific progress and to the reproduction of the nation's long-standing cultural heritage, setting in motion a dialogue between empirical data and philosophical reflection that endures to this day.

Spirit has deep roots in French intellectual history. Its champions called themselves "spiritualists," and they held that selfhood is irreducible to matter alone. Since the seventeenth century, French spiritualism had been committed to the existence of a subjective dimension of reality accessed via inner experience.[2] It has been suggested that this tradition was "the most French of all philosophical orientations."[3] Consider the Enlightenment poet Louis Racine (1692–1763), who gave voice to spirit's superiority over matter:

> I think, that glorious light that guides my tongue,
> My every motion, ne'er from matter sprung:
> I glimpse my greatness. This unwieldy frame
> Is not the all I am, the all I claim.[4]

The idea that selfhood involves more than the body was foundational to spiritualist thought, a historically expansive and theoretically coherent movement whose story has yet to be told in the English-speaking world. If the term *spiritualism* might embarrass us today, that is because our intellectual climate is no longer one in which the idea of spirit compels conviction as it once had. For women and men living before the twentieth century, spirit was not a spooky apparition. It was a constitutive feature of human life. If we relegate spirit to history's dustbin of dead concepts, then we are liable not just to forget the past. We risk abandoning a vital resource for bringing the sciences and humanities into dialogue today. As I argue in the following chapters, support for the cause of spiritualism—and for its formidable legacy in French history—reached a crescendo at the dawn of the twentieth century thanks to the country's investment in neurology and psychology.

Disputes over the nature of self-knowledge turned on the spiritual and the material: which fields (biological or philosophical; neurological or cultural) conferred meaning and purpose on people's lives? At the beginning of the twentieth century, when the disciplinary cleavages now easily taken for granted were only beginning to congeal, spiritualism came to the fore as the leading voice for an alternative epistemological configuration. Henri Bergson, the movement's most celebrated proponent, articulated this vision in a 1901 speech before the French Society of Philosophy. His immensely popular books made the case that freedom is temporal; it inheres in our nature to change incessantly and is, therefore, completely unlike material objects in space. Reflecting on the motivations for his defense of free will, Bergson said, "I initially considered the manifestations of matter not in their simplest forms, in physical facts, but in their most complex forms, in physiological facts. And it was not physiological facts in general that I focused upon, but cerebral facts."[5] Bergson belonged to a wave of thinkers who decided to engage rigorously with shifting concepts of matter in order to move beyond them. "By restricting spiritualism to these extremely narrow boundaries, it seemed to me that we could indefinitely increase its fertility and its force, making it accountable to those who reject it, bringing to it a theory of knowledge through which it could dissipate its obscurities, and finally to make it the most empirical of doctrines in terms of its methods, and the most metaphysical in terms of its results."[6]

Why did French thinkers—Bergson chief among them—decide to cull from psychology and neurology conceptual resources with which to uphold human autonomy and personal creativity? How did a tradition as seemingly démodé as spiritualism—with its dualist commitment to the existence of material and immaterial realities—enter into partnership with some of the most innovative scientific breakthroughs of the era? What might this history offer to those of us seeking to bridge scientific and humanistic knowledge today? These questions guide the following chapters. In telling the story of French spiritualism, I will show how this oft-forgotten but impressively relevant intellectual formation revolutionized the metaphysical and theological study of the self on the basis of the very sciences that appeared to cast doubt on the authority of philosophy and religion.

Another one of my aims is to enrich genealogies of selfhood in Europe.[7] When we give an account of what at bottom we are, our narratives rely on categories that are not altogether of our choosing. They depend on a contingent time and place. The sources that made interiority—that is, one's reflexive relationship to experience—into a defining feature of our existential condition have been a fruitful and ongoing area of historical exploration.[8] According to Charles Taylor, the modern self is characterized by "disengagement."[9] We assume an epistemological stance set over and against the world, sustained by the dyadic relation between subject and object. On the one hand, objectivity constitutes a rule-bound realm whose properties and relations are items of empirical knowledge. On the other hand, subjectivity recedes into the realm of thoughts, feelings, and desires. The thinkers in this book were quintessentially modern in so far as ideas of spirit presupposed the duality between interiority and exteriority. These concepts, however, were thrown into disarray when the cerebral center inside the self was shown to be no different than the physical matter outside it. Spiritualists' response helps us to make sense of the complex contestations over the structure and composition of interiority, committed as these thinkers were to defining the affective density and conceptual integrity of embodied experience.

The debates over selfhood threaded through the following chapters offer a timely reminder that the mind-body problem is neither universal nor timeless. Although it has been among the most pressing and persistent problems in Western thought, answers prove elusive because the very problem is hardly innocent. Its terms reflect the tangled dimensions of our human form of life. How the "conscious," "intentional," or "cultural" relate to the "physical," "biological," or "natural" hinges on the joints along which we cut these dimensions of the self. They assume different shapes, respond to various pressures, and serve distinct purposes in particular social and political contexts at contingent moments in time. What I hope readers take away from the story of spirit and matter in modern France is a critical appreciation that the mind-body problem is not exclusively scientific or philosophical. It is also historical.

The Legacies of Spirit in France

The origin of French ideas of spirit can be traced to Michel de Montaigne (1533–92). His essays argued that knowledge ultimately has its source in oneself. Since life is in constant flux, it is imperative, Montaigne held, to look inward, control one's passions, seek happiness, and accept that absolute certainty is unattainable. Montaigne's idea that introspection confers a privileged kind of knowledge found a systematic account in the *cogito* of René Descartes (1596–1650). The clear and distinct perception of thought, he claimed, reveals immediate knowledge distinct from knowledge mediated by observation of the external world. This epistemological distinction between two forms of knowledge depended on Descartes's metaphysical division between two realms: a thinking substance, *res cogitans*, and an extended substance, *res extensa*. Cartesian dualism served to safeguard the veracity of rational knowledge from any attempt to derive its origins from the material world. The dualism between the truths within us and those outside us found expression in the twin sources of knowledge that Pascal (1632–62) posited in his *Pensées* (1670), emblematized by the famous dictum "We know the truth not only by reason, but also by the heart."[10] The division between spirit and matter was a cornerstone of eighteenth-century French thought. Jean-Jacques Rousseau (1712–78) explored the spiritual bases of freedom in his meditative and political writings. He set humans' capacity to control their own life in opposition to the regular relations of matter and motion. "Nature commands all animals, and the beasts obey. Man receives the same impulsion, but he recognizes himself as being free to acquiesce or resist; and it is above all in this consciousness of his freedom that the spirituality of his soul reveals itself, for physics explains in a certain way the mechanism of the senses and the formation of ideas, but in the power to will, or rather to choose, and in the feeling of that power, we see pure spiritual activity, of which the laws of mechanics can explain nothing."[11]

Whereas Rousseau took spirit to be the engine of freedom anterior to the determinate order of nature, materialists denied any metaphysical separation. Claude-Adrien Helvétius (1715–71) claimed in *Of Spirit* (1758) that physical sensibility is "the only quality essential to the nature of man."[12] Denis Diderot (1713–84) famously described living spirit in *D'Alembert's Dream* (1769) as an assemblage of mobile matter. The most comprehensive system of materialist thought came in what Étienne Bonnot de Condillac (1714–80) called sensationalism. His thought experiment of an insentient statue, which gradually acquired all its ideas through sensations of the material world, represented the zenith of anti-spiritualist thought. It catalyzed the materialist impulses of the nineteenth-century human sciences, which considered the reduction of phenomena to their efficient causes to be the hallmark of sound scientific explanation. Emblematic was the French physiologist Pierre Jean-George Cabanis (1757–1808), who argued, "The brain secretes thought like the liver secretes bile."[13]

The nineteenth century witnessed the intensification and consolidation of an intellectual lineage known as French spiritualism.[14] It took psychology to be philosophy's raison d'être. The work of Marie-François-Pierre Gonthier de Biran (1766–1824), or Maine de Biran, was foundational. He argued that the activity of thought, guided by inward reflection, is part of the very fabric of the world. Our individual viewpoint is not an auxiliary perspective on reality; it is the world viewed from the inside. His writings were critical to the development of psychology in that Biran demonstrated how spirit constituted neither a stable point nor a substance within the self but a dynamic activity connecting the individual mind with the world of which it is a piece. The activity at the core of Biran's psychology was the feeling of motor effort. In *The Influence of Habit on the Faculty of Thinking* (1802), he explained how repeated bodily activity generates an immediate impression of selfhood. Lively sensations of one's own muscular exertion—and not the passive sensations received from the external world—constitute the unique source of self-knowledge.

The majority of Biran's writings on the psychology of fleshly volition came to be known belatedly thanks to his heir Victor Cousin (1792–1867). Cousin's brand of spiritualism—known as eclectic spiritualism—was an official state ideology under the July Monarchy. The period was marked by liberal reforms that expanded popular sovereignty yet subdued egalitarian movements for social change. From 1830 to 1851, Cousin held lofty positions in the Ministry of Instruction and the University of Paris. With his power, he enshrined psychology as the centerpiece of the official philosophy curriculum in secondary education (lycées). Education became an apparatus for reproducing a state-sanctioned elite. Students studied the march of Western ideas culminating in the nation's crowning achievement: the rational principles of spirit. Like Biran, Cousin took psychology to be a vehicle for his Christian faith. The psyche was no mere accident in world history; it was endowed by God and touched the absolute.

Numerous historians have brought attention to the politics of Cousin's spiritualist edifice, which sought to bring about a *juste milieu* that would restore order to France in the aftermath of the 1789 revolution.[15] According to Jan Goldstein, Cousin's disciples "constructed their psychology around an immaterial self, or *moi*, that (they insisted) was given to its possessor whole and a priori."[16] The driving question of the nineteenth-century curriculum was the meaning and nature of the person. Under Cousin's authority, the exclusively philosophical answer secured a stable ground of knowledge. An anti-naturalist approach to psychology transcended the vicissitudes of actual scientific research. Instilled in elite students privileged to enroll in lycées, the enlightened values of aristocratic authority reigned in the chaotic passions of popular democracy.

Although spiritualism's pedagogical-political power has been documented as a defining feature of the institutional and intellectual environment of modern France, historians have largely neglected the significant tensions *within* ideas of spirit.[17] Resistance to Cousin's authority brewed under the Second Empire of Napoléon III (1852–70). Many philosophy instructors sought to revive the corporeal psychology of Maine de Biran, whose writings brokered a productive exchange with the human and life sciences. An intellectual metamorphosis was afoot. In a book titled *The New Spiritualism*, and addressed to "us free disciples of Victor Cousin,"[18] Étienne Vacherot pursued a different philosophical direction: "It no longer seems possible to maintain a dualism between two substances such as the Cartesian philosophy believed to have solidly established on the basis of consciousness. Experimentation has revealed the constant correspondence between psychic phenomena and physiological phenomena. It is on the way to showing that the correspondence is not any less absolute between the organ of the brain and the faculties of the soul: to ensure that it is no longer possible to conceive the two principles as two activities in a state of separation."[19]

How did it come to be that such a spiritualist stalwart who ascended to the Sorbonne as an apprentice of Cousin would find in the brain sciences the resources to integrate the somatic and spiritual dimensions of selfhood? Vacherot had previously argued that they were rigidly opposed in *Metaphysics and Science* (1858). His about-face in the modernizing social climate of France was symptomatic of a philosophical wave engulfing the country.

The shifting tides of spirit flowed into university classrooms. When Jules Lachelier (1832–1918), a philosophy professor at the prestigious École normale supérieure, opened his course on psychology in 1866, he asked the students to identify their object of study: the soul, the brain, or the self? He immediately eliminated the first two choices. The soul represented an abstract being that could not be known with certainty, and studying the brain, he advised, would dissolve our spiritual capacities into matter. Rather, Lachelier reasoned that "psychology stakes out the self: the given principles of the science."[20] In accord with the French academic establishment, philosophical anthropology—the study of human beings' nature—was the course's guiding theme. Neither an abstract substance nor a material entity, the self, Lachelier claimed, constituted the nexus of human experience. What made the study of the self *scientific* was what he called psychology's "subjective method"—a practice that reflects inward, traces our dynamic mental activity, and ascertains its seat in the life of spirit. Lachelier spent the year laying out the science of spirit. For the students enrolled in the course—an elite cadre of young bourgeois men—the lessons no longer served just to edify their minds. The curriculum equipped them with the tools to bring France's intellectual heritage to bear on psychologists' bourgeoning research in neurophysiology.

After he finished the course, a young Émile Boutroux (1845–1921) departed for Heidelberg. Like many French students who came before him,

Boutroux left to study in the bastions of German philosophy where Immanuel Kant's legacy endured. But his time was cut short. The outbreak of the Franco-Prussian War in 1870 tore at the entwined intellectual legacies of the two countries. "Since the Prussians have been on our land," Lachelier wrote to his student, "they have proved quite well their energy, their spirit of order and foresight, as well as that self-righteousness that you noticed in Heidelberg."[21] While Boutroux complained that his Catholic sensibilities were out of place among the Protestant faculty, Lachelier narrowly escaped the German bombardment of Paris's Left Bank. He took up a post in the National Guard to defend the Empire of Napoléon III.[22] The regime swiftly fell. And the revolutionary Paris Commune that sought to take its place was soon violently squashed by the French army.

When Boutroux returned, he found his homeland humiliated. Many attributed France's defeat to the Germans' scientific progress and—above all—to the derelict state of French research institutions. Regime change ensued. The French Third Republic was born at the height of war and amid intense political turbulence. Only when the prospect of a return to royalism subsided with the elections of 1877, and Republicans won a parliamentary majority in 1879, did the government commence in earnest with the modernizing reforms that had inspired its inception. The education system was overhauled to remake the nation.[23] Primary schools expanded, and lycées set about cultivating an ostensibly meritocratic citizenry. Nobility would not determine social or political advancement. Most, however, could not access lycées, let alone universities. (The Third Republic entertained few aspirations to combat actual social hierarchies so long as formal meritocracy prevailed—in principle—for all.)[24] The youth of the bourgeoisie who did enter lycées learned under civil servants who belonged to what the Republican politician Léon Gambetta (1838–82) heralded as the new social stratum. Philosophy instruction came to serve a different purpose than it had under Cousin's authority. Lessons alloyed social renewal to scientific advancement. A new generation of professors were charged with disseminating the official curriculum in the provinces beyond Paris. Boutroux was sent to teach at Lycée Caen, a northern town situated near the English Channel, where he inculcated the spiritualist methods that he had learned under Lachelier's tutelage alongside scientific and secular values.

Like his peers, Boutroux carried out his teaching responsibilities while writing his doctoral thesis. In *The Contingency of the Laws of Nature* (1874), Boutroux argued that the materialist pretensions of the natural sciences—and neurological inquiry in particular—were anything but. He demonstrated that "physiological phenomena are not absolutely determined, but that they contain a radical contingency."[25] In other words, spirit was a lively activity in the natural world, imparting freedom to physical processes as well as to cerebral matter. In the thesis that he wrote in the early years of France's precarious democracy, Boutroux showed that the innovations transforming psychology into a physiological science—one that took nervous action and not the disembodied self as its object—had breathed new vitality into spirit.

Boutroux and his peers did not take up the defense of spirit as their teachers had before them. They were not content to police the boundaries of selfhood for fear that clinical, quantitative, and experimental methods would threaten philosophers' metaphysical purview. Prior generations of spiritualist thinkers allied to Cousin had dismissed—if not ignored—scientific research. Why did a new wave of thinkers decide to go through the brain sciences carefully and rigorously in order to reinvent spirit?

The following chapters document this arresting transformation in French intellectual history. My objective is to make sense of the myriad names that appeared for the changes underway. Lachelier dedicated his lectures to "spiritualist realism."[26] Allusions to "the new spiritualism" were abounding.[27] Paul Janet (1823–99) claimed that the old spiritualism was simple: "Accept God, the soul, liberty, and the future life."[28] Younger figures reconceived these principles within the sensory-motor lexicon of neurophysiology. The new spiritualism "is not a new doctrine," one author wrote in 1884; "it is spiritualism renewed by science."[29] The panoply of terms testifies to the exigency driving thinkers to self-consciously distinguish the new spirit from the old. The characterization of this transformation as a turn to materialism took hold thanks to a critic of the movement. A defender of the old guard decried what he saw as its abnegation in the form of "neo-materialism." As for its commitments, "these three notions are combined and gathered: fundamental contingency, unlimited becoming, internal life anterior to intelligence and intelligibility—creator of one and the other; with them we end up with the product—the new philosophy—which represents the exact antipode of rationalism."[30] A materialist moment inflected the spiritualist movement by the turn of the century.

The impulse to draw from the brain sciences the very means to go beyond their results was shared among the cast of characters in this book. The protagonist is Henri Bergson. He has been celebrated as a luminary thinker ever since holding a chair of philosophy at the Collège de France from 1900 to 1921. Bergson's 1927 Nobel Prize in Literature cemented his position in the canon of European thought. The thrust of his oeuvre, I argue, was to steer a materialized spiritualism into the twentieth century. In his opus of 1907, *Creative Evolution*, Bergson made clear where revamped ideas of spirit broke with those of the past: "Philosophy introduces us . . . into the spiritual life. And it shows us at the same time the relation of the life of the spirit to that of the body. The great error of the doctrines of the spirit has been the idea that by isolating the spiritual life from all the rest, by suspending it in space as high as possible above the earth, they were placing it beyond attack, as if they were not thereby simply exposing it to be taken as an effect of mirage! . . . They are right to believe in the absolute reality of the person and in his independence toward matter; but science is there, which shows the interdependence of conscious life and cerebral activity."[31]

Despite Bergson's fame, his ideas were not sui generis. An argument of mine is that Bergson became the most articulate spokesman for a wider intellectual formation in modern France. In this context, we can better comprehend the theoretical stakes and mass appeal of his compendium of concepts: the duration of lived experience; philosophical intuition; the planes of consciousness; the vital impulse (*élan vital*). Their global transmission spanned both sides of the Atlantic.[32] Scholars have also traced Bergson's reception beyond Europe and America.[33] Notwithstanding his eloquence as a writer and speaker, Bergson's methods were not entirely his own. The numerous thinkers who play supporting roles in this book also found in the science of matter a philosophical support to bring our spiritual powers into connection with the body.

These included Lachelier and Boutroux. There were others: Félix Ravaisson (1813–1900) drew on biology to elucidate the corporeal dynamics of habit; Alfred Fouillée (1838–1912), the philosopher of *idées-forces*, composed a magisterial corpus excavating the spiritualist facets of evolution, psychology, and sociology; Jean-Marie Guyau (1854–88), France's "literary Nietzsche," extracted a protean theory of the self from experimental psychology; Maurice Blondel (1861–1949), the Catholic pragmatist, rejuvenated faith using developments from physiological psychology; the philosopher and mathematician Édouard Le Roy (1870–1954) traced the genesis of scientific systems back to our embodied interactions with the environment.

This constellation of thinkers contributed to what I call "spiritualist materialism." The term is my own. It brings together a capacious movement characterized elsewhere as the "golden age" of spiritualism.[34] But the overlapping commitments that lent it conceptual consistency have yet to be examined. Spiritualists worked from within a gathering materialist account of consciousness to advance a vision of matter other than that held by progressive voices of the nineteenth century who sought to reduce selfhood to the body's nervous pathways. Emblematic was the German physiologist Carl Vogt (1817–95), who declared, "thoughts stand in the same relation to the brain as gall does to the liver or urine to the kidneys."[35] Karl Marx's (1818–83) account of the material foundations of history relied on a similar approach to matter, as when he famously remarked, "The tradition of all dead generations weighs like a nightmare on the brains of the living."[36] It is important to point out that spiritualists did not constitute a monolithic school. Nonetheless, they shared more than a common sensibility—more, to be precise, than an ambition to occupy a "middle ground" between the idea of a freely willing subject and biological determinism.[37] Among those for whom spiritualist materialism held sway, we find of a countervailing approach to matter: one that sought to inflate our creative powers using brain studies.[38] These became the raw material for doing philosophy. The figures whose story I tell in this book are worth revisiting today because they were incisive readers; they interpreted the scientific literature of their time in an effort to make spirit matter.

Intellectual Itineraries in the Fin de Siècle

The end of the nineteenth century was an era of immense intellectual production in Europe. Artistic expression blossomed; philosophical creativity flourished. Technology laid the conditions of accelerated productivity: electricity prolonged the workday; railroads and telephones brought distant people in contact; the diminishing cost of print facilitated journals' international circulation across the continent. Scientific knowledge became more widely available than ever before, and optimism in its unbounded potential thrived.

The natural sciences expanded their reach. In their taking the world to operate like a machine, the principles of matter and motion likened human functions to clockwork. In France, faith in science went by the name of positivism. In his *Course in Positive Philosophy* (1830–42), Auguste Comte (1798–1857) had argued that humanity evolved across three stages: theological, metaphysical, and positive. His broad following included polymaths such as Émile Littré (1801–81), Ernest Renan (1823–92), and Hippolyte Taine (1828–93), who espoused the virtues of scientific inquiry, which would uncover the observable realities—that is, positive facts—underlying all phenomena, and thereby shepherd society to maturity. Religion would be relinquished, along with its vestigial residue in the form of metaphysics. Across western Europe, a culture of scientism celebrated the growth of ever-larger compendia of empirical data. The incipient brain sciences were imbricated with other developments—thermodynamics, electromagnetism, and cell theory—believed to augur the idea that everything, including the self and society, could be mastered with predictive certainty. European men and women living at the turn of the century grappled with a rationalized world dominated by ironclad laws.

According to a long-standing historical narrative, scientific rationality's encroachment into the depths of the mind provoked an intellectual revolt that emphasized the irrational character of human thought and action.[39] Emotion, sentiment, and instinct took center stage. They were shown to elude explanatory models originally developed for mathematics, physics, and chemistry. Many were anxious to safeguard human freedom from causal explanation when the limits of the natural sciences were exposed, and a crisis in their very authority ensued. Titanic thinkers across western Europe—including Henri Bergson, Benedetto Croce, Wilhelm Dilthey, Sigmund Freud, Carl Jung, and Friedrich Nietzsche—carved out alternative models to explain the unruly nature of selfhood, thus laying the foundations of modern social theory. The historian H. Stuart Hughes had inaugurated this narrative when he described fin de siècle thought as a "revolt against positivism."[40] Although over half a century has transpired since, it is hard to underestimate just how widespread the narrative remains among historians of Europe who characterize the modernist impulses of the period in the terms of a "revolt against mechanism,"[41] a "reaction against materialism,"[42] a "rejection of positivism,"[43] or a "revolt against rationality."[44]

This book's story is different. It is about a moment in European thought motivated neither by moral anxiety in a disenchanted world, nor by aesthetic distaste for bourgeois society, nor even by apprehension of the natural sciences' overreach. The story of French spiritualism is about a project to elicit from the scientific breakthroughs of the late nineteenth century more expansive notions of rationality, positivism, and materialism. The thinkers in this book radically dilated the bounds of scientific facts to include the data of subjective experience. Ravaisson called this "spiritualist realism or positivism."[45] Bergson was committed to "positive metaphysics."[46] Le Roy upheld "spiritualist positivism," which, "far from having been called from outside, as it were, by metaphysical and moral preoccupations . . . has appeared from the inside of science, under the pressure of its internal needs, and in contact with its very facts and theories."[47] Conceptual resources within neurology and psychology fueled an expansion of the positive. Indeed, the materialist moment in spiritualism casts into stark relief that a wide swath of fin de siècle culture was propelled by an abiding sense of admiration for the results of empirical inquiry—and not by an oppositional stance against the scientific community.

A narrative more befitting the intellectual history of the period would emphasize the sciences' conceptual surplus. The spiritual and material vexed neurophysiology to the point that any explanation for their connections exceeded scientists' working categories. That task was left over in the form of a metaphysical remainder.

The French scientific establishment recognized that "as a science, psychology takes the human spirit as its object."[48] Its fraught connections with the body's materiality had persisted since the nervous system's functions were discovered at the beginning of the nineteenth century. Sensory nerves conducting signals from the skin were shown to connect to the spinal cord by the posterior roots while motor nerves conducting impulses through the muscles depart from the anterior roots. Thanks to independent studies by Charles Bell (1774–1842) in England and the French physician François Magendie (1783–1855), what came to be known as the Bell-Magendie law offered an analytical framework to decompose the mind's immaterial faculties into sensory-motor categories of the body. Mental processes were thus made amenable to experimentation. A key area of exploration was the reflex: the nerves' automatic movement in response to stimuli. The British physiologist Marshall Hall (1790–1857) noticed that decapitated animals often continued to move. When he irritated a headless newt, its tail contracted. Reflex action explained the animal's involuntary activity, which did not depend on the brain's active coordination. Instead, Hall contended that reflex activity depended on the "spinal system."[49] Because no spirit was thought to be involved, the reflex arc opened the possibility of harmonizing the neural pathways' sensory-motor circuitry with matter and motion. But metaphysical notions could not be so easily eliminated. Ruth Leys argues that dualism was reinscribed within the brain sciences because "the central nervous system was subject to the laws of the reflex as high as the medulla but no higher: above the cord was the entirely different cerebral system, the seat of the immortal soul."[50] Subsequently, the German physiologist Johannes Müller (1801–58) extended reflex activity beyond the medulla oblongata, to the entirety of the cerebral structure. He demonstrated that sensation depended on the brain, which coordinated overlapping muscle groups. His conception of the reflex was of a piece with Hall's dualism; both claimed that the spinal cord did not initiate muscular movement. Herein lied the motivation for distinguishing the functions of the spine from those of the cerebrum. The former were sensory-motor. The latter, as the British physiologist William Carpenter (1813–85) claimed, were "restricted to *intellectual* operations; understanding, by that term, the operations which are concerned in the formation of a voluntary determination."[51] The brain housed the will. But its mandates were carried out by muscular contractions descending from the cerebellum down through the spinal cord. This meant that the nervous system was the servant (but not the source) of agency. Despite immense strides, the science of neural mechanisms gave rise to conceptual dilemmas that surpassed the explanatory frameworks available.

Far easier was it to divide the spiritual and the material than to bring them together.

The history of ideas of spirit opens a window onto an alternative genealogy of the neurosciences. The cleavage between intellectual and sensory-motor capacities revealed the mind-body problem to be a constitutive feature of neurophysiology. The problem came to a head in new research programs during the final decades of the nineteenth century. The quantification of sensory states, the measurement of nervous action, and the identification of mental operations in distinct brain regions (i.e., cerebral localization) struggled to suture self and soma. Instead, they were torn asunder. Consequently, exploration into the structure and functions of the nervous system bled onto metaphysical terrain that offered fertile ground to submit the problems within neurophysiology to an elevated level of reflection. There was a higher dualism at the heart of the self. The spiritual components of corporeal experience involved not just the autonomous power of thought (*cogito*), as thinkers since Descartes had conceptualized our rational capacities. Equally fundamental was the embodied exercise of volition (*volō*).[52] Moving the body was shown to make spiritual powers enter and transform one's biological and social milieu. Subjectivity was riven. Between thinking and willing, the self was reimagined as *Homo duplex*.

The commitments and consequences of brain research came to light, I am suggesting, as knowledge about the organ refracted through the prism of spirit. In turn, the metastases of spiritualist concepts resulted from the problems generated by the brain sciences. By tracing a history of problems across the intellectual and institutional landscape of modern France, the ensuing chapters lay bare the conceptual and cultural resonances of emergent brain research.[53]

In the first chapter, I trace the disjointed trajectories of spiritualism beginning with Maine de Biran's writings. Although he published little in his lifetime, the posthumous publication of Biran's embodied psychology came to serve as a post factum origin for spiritualist materialism in the late nineteenth century. I take up this transformation in the second chapter, which explores the rise of quantitative psychology. In calculating the intensity and speed of neural activities, psychophysics and mental chronometry were productive dialogical partners in the spiritualist project to articulate qualitative facts of first-person experience that were rigorous enough to complement numerical data. Debates over what counted as a psychological fact resonated across French society. But they were felt acutely in lycée classrooms, the subject of the third chapter. I examine textbooks and course notes, particularly those written by students of Bergson. He became the most successful representative of the materialist turn in spiritualism thanks in large part to his mastery of materials in the state-mandated philosophy course. In the fourth chapter, I explore his peers who studied psychopathology and cerebral localization. These thinkers' interpretations were not always faithful to scientists' understandings of their own work. But they embraced a critical perspective that allowed them to see what scientists often did not—namely, that shifting concepts of neural matter entailed more dynamic notions of embodied experience, including the contributions of habit, exhaustion, and memory to thought.

This story about science and selfhood in the fin de siècle will, I hope, leave readers with a fresh interpretation of the broader contours of French intellectual history. For too long, spiritualism has served as a facile pivot point for historians' oppositional narratives.[54] Characteristically, Michel Foucault noted a fissure in French philosophy when he reflected on the past two centuries:

> It is the one that separates a philosophy of experience, of meaning, of the subject, and a philosophy of knowledge, or rationality, and of the concept. On one side, a filiation which is that of Jean-Paul Sartre and Maurice Merleau-Ponty; and the other, which is that of Jean Cavaillès, Gaston Bachelard, Alexandre Koyré, and Canguilhem. Doubtless this cleavage comes from afar, and one could trace it back through the nineteenth century: Henri Bergson and Henri Poincaré, Jules Lachelier and Louis Couturat, Pierre Maine de Biran and Auguste Comte. And, in any case, it was so well established in the twentieth century that, through it, phenomenology was admitted into France.[55]

It has been debated whether Foucault's division between a philosophy of experience and a philosophy of the concept adequately captures the nuances of French thought.[56] The division is nonetheless symptomatic of an enduring misunderstanding that those who were preoccupied by the existential structure of consciousness—as the thinkers in the book certainly were—did not also have deep questions and concerns about scientific concepts. Not only do our accounts of French history suffer as a result. We also lose sight of the playful uses of scientific literature: its malleability in the hands of incisive interpreters from disciplines outside the sciences.

By no means motivated by a revolt against materialism or positivism, spiritualist materialism was actually sustained by a conflict with French neo-Kantianism. As I argue in the fifth chapter, Kant's self-proclaimed inheritors in France (including Léon Brunschvicg, Baptiste Jacob, and Charles

Renouvier) sought to safeguard the self's rational capacities from psychology and neurology. Inspired by Kant's three *Critiques*, a gulf separated the social realm of norms issued by the intellect from the natural realm of causes analyzed by the sciences. Spiritualists eschewed this picture of selfhood and advanced another: that thinking reaches out through the body and interacts with the material around us. The contingent penetrated the causal. In delineating the multiple dimensions of experience, these thinkers also responded to the fate of religion amid the secularization of Europe. As I show in chapter 6, spiritualist materialism stoked the modernist crisis in Roman Catholicism at the turn of the twentieth century. Christianity had always been inseparable from French spiritualism. What we find during this period was a profound reimagining of scripture, divinity, and dogma on the basis of our embodied dynamics. Faith came to be seen as a bodily practice. As a result, spiritualist thinkers drew the Vatican's ire. Curiously, the work of Bergson—a Jew—proved to be instrumental in melding the magisteria of religion and science.

From our vantage point, the science of the nineteenth and early twentieth centuries might seem primitive. We now know that the nervous system is composed of neurons, each of which has a cell body with one long projection called an *axon* and numerous shorter *dendrites*. We also know that neurons communicate with each other via *synapses*, which receive and release chemical molecules called *neurotransmitters*. Judged by our standards, past scientists' neurophysiology was incomplete, dependent as it was on a mechanical picture of the brain that was not privy to the biochemical insights that came about later in the twentieth century. Yet, their knowledge gave rise to the conceptual problems that continue to vex our neurosciences today. It's not implausible to say that we know more about the solar system than we do about the nervous system. To grow our knowledge demands not just that we accumulate more biological data but also that we grapple with foundational questions of selfhood in which the brain plays a part.

The thinkers in this book did that not just by writing and reading. Driven by the pressing problems of their day, they also gathered and worked in spaces such as classrooms, laboratories, journals, societies, and libraries. In these concrete institutional networks, men (and admittedly few women) brought spiritualist ideas into confrontation with the scientific, political, and cultural tumults of European society. They did not always offer correct answers. Then again, my aspiration is not to vindicate them (even though an undeniable affinity nourished over years of dwelling with their manuscripts, letters, marginalia, library records, and diaries will shine through). I instead want to show how the materialist moment in spiritualism disclosed new possibilities. Indeed, I see this book as a modest excavation of past possibilities worth revisiting, especially today amid the dissilient advances of psychology and the neurosciences: namely, that investigations into the body do not only offer *solutions* to our questions about the nature of selfhood, but also harvest interpretive *problems* that lend renewed traction to inquiries and methods beyond the frontiers of science.

✹ CHAPTER 1 ✹

The Formations of French Spiritualism

The Universal Exposition of 1867 took place in Paris. The French emperor, Napoléon III, seized the opportunity to put the nation's scientific prowess on display before a global audience. Nine million visitors came to the military grounds of the Champ de Mars, the site of the Exposition Palace. Fairgoers passed through the concentric rings of the large oval edifice, looking in awe at industrial inventions such as the first elevator, a model of France's Suez Canal (still under construction), and a towering fifty-ton Krupp cannon that the Prussians would use three years later to pummel the Second French Empire in the War of 1870. Passing to the inner rings revealed the fine arts exhibition in the middle of the building.[1] On display were sculptures, cabinetry, dresses, and paintings. The architectural configuration was not random. In moving from periphery to center, fairgoers passed from science to art. France's spiritual achievements were, both literally and figuratively, the centerpiece of the 1867 Exposition.

For the event, the imperial government commissioned a *Report on Philosophy in France during the Nineteenth Century*. It was the idea of Victor Duruy (1811–94), the minister of public instruction. His goal was to boast the nation's intellectual heritage and to forecast its future. To write the report, Duruy recruited Félix Ravaisson, a little-known philosopher who soon would have a wide influence over the French cultural landscape. He traced the march of Western philosophy bequeathed to France, from antiquity to the scholastics of the Middle Ages, through the great polymaths Descartes, Malebranche, Condillac, and ultimately the philosopher Maine de Biran. Ravaisson's aim was to the champion his country's intellectual legacy, the idea of spirit: *spiritus intus alit*.[2]

The report framed the spiritualist heritage in opposition to the insurgent materialisms of the mid-nineteenth century. Hippolyte Taine's materialism drew wide support in France. His writings synthesized art, history, and the psyche according to causal laws of matter and motion. The various fields were harmonious, he believed, with electrical impulses traversing the human nervous system. It followed that the achievements of human knowledge and sensibility could be derived, at bottom, from the species' physiology. On the occasion of the Universal Exposition, Taine published a rival report in which he claimed that the central fissure in the history of French philosophy ran between positivism and spiritualism.[3] The former was grounded in the authority of science. By contrast, "all the difficulty [of spirit] consists in preserving for itself the optical illusion that makes us take causes as beings, that transforms metaphors into substances, and that gives consistency and solidity to phantoms."[4]

Ravaisson traced a different trajectory. His 1867 report foretold the emergence of a third avenue beyond spiritualism and positivism: what he called "spiritualist realism or positivism": "The signs are in place; we can foresee a philosophical epoch soon to come whose general character will be the predominance of what might be called a *spiritualist realism or positivism*, having for its motivating principle the awareness that consciousness has of itself, a self understood to be an existence from which all other forms of existence derive and depend, and which is nothing other than its own activity."[5]

This dynamic and reflexive conception of the self did not stand apart from the material world. According to Ravaisson, the body was imbricated with "a fully active, and thus fully spiritual nature, its existence complete and absolute; a nature from which it follows that thought, will, and love are one and the same thing, a flame, as it were, without material support, and thriving on itself."[6] His "spiritualist realism or positivism" parted with Taine's reductionist interpretation of science. Ravaisson eschewed the ideology that physics' causal models could eliminate God and consciousness from nature. Rather, divinity and spirit traversed the natural world.

"Generations of students," Henri Bergson later reflected, "have learned [these pages] by heart."[7] For French philosophers who came of age in the second half of the nineteenth century, Ravaisson made it known "that the serious study of the phenomena of life must lead positive science to widen its framework, and go beyond the pure mechanism in which it has been enclosed for the last three centuries."[8] Writing as the official mouthpiece of French education, Ravaisson encouraged fellow philosophers to study the inner contours of subjective experience in conversation with the biological data of natural science.

This chapter traces the genesis of "spiritualist realism or positivism," a movement that brought together diverse thinkers across the nineteenth century. Many rallied around Ravaisson's report, imagining a horizon for ideas of spirit by working backward and forward in time. These figures reinvented their philosophical tradition while advancing a new vision for selfhood, a vision that gave prominence to the creative and productive forces of nature within us.

The report was a bellwether. For much of the century, Victor Cousin's eclectic spiritualism had dominated the academic establishment, where science and philosophy were configured as independent magisteria. There could be no bridge between them. Since he began lecturing at the École normale supérieure in 1815, Cousin had promoted a historically informed curriculum as the backbone of psychology. Turning away from the external world, psychology was "thought folding back upon itself, and contemplating the spectacle presented by itself."[9] According to Cousin, introspection revealed the self to be an immaterial substance that anchored the rational principles by which the world appeared before thought. Up through his final treatise, *The True, the Beautiful, and the Good* (1853), Cousin treated psychology as a historical gateway "to subordinate feeling to the spirit, and to aim at raising up and enlarging man by all the means that reason admits."[10] This method was *eclectic*, as Donald Kelley highlights, because "history in effect took precedence over unassisted and unencumbered reason and became 'first philosophy.'"[11] Students studied the canon of Western philosophy in order to bring their mental faculties into moral alignment. They would embody the ideal citizen, which Jan Goldstein argues was integral to the bourgeois masculine ideology of post-Revolutionary France. Her book on the subject shines a light on "the hegemony of Cousinianism in France, its ability to beat out its competitors and impose its concept of the self on a significant segment of the population."[12] Indeed, Cousin's pedagogical-political power dominated the French academic establishment well into the second half of the century.

Following Cousin's death in 1867, Ravaisson stated his opposition in no uncertain terms: "In these last years, eclecticism, although still in possession of nearly all public education, has lost its credit and influence."[13] The report gave voice to thinkers who were frustrated by Cousin's enmity toward science. Ravaisson had been summoned numerous times to write official accounts on the state of academic institutions.[14] When it came time to survey French philosophy, many cautioned him to announce his break with Cousin delicately.[15] But he went further once cracks emerged in Cousin's authority. The report gestured toward a post-eclectic future for spirit.

The report was presented before the Senate at a time when resistance to the Second Empire had gained traction. Duruy read the report aloud, enunciating the stakes confronting France: "spiritualism is the raison d'être of the University, it is found in the lycées, it is in the faculties, which is to say higher education"; conversely, "the threatening progress and invasion" of materialism "is everywhere in retreat."[16] In suit with Ravaisson, Duruy exalted science for laying the foundations of spirit: "Let science go, let it do its work: knowledge of the soul lies at the end."[17]

His remarks were consequential. Academic philosophy was at a turning point. The philosophy course in secondary education—mandatory throughout the nineteenth century—had been suppressed in 1852 following the coup of Napoléon III.[18] The emperor censored philosophy (alongside history), claiming they were anticlerical disciplines that would nourish radical resistance. When Duruy took over the Ministry of Public Instruction in 1863, he put in place liberal educational reforms, reopened the philosophy course, and begrudgingly enforced eclectic spiritualism as the official doctrine. By the time of the 1867 Exposition, however, a new spiritualism was afoot. Many philosophers expressed distaste for Cousin's eclectic edifice.

The report offered a blueprint for a materialist turn in ideas of spirit. According to Ravaisson, the spiritualist tradition split in two directions: the official psychology of Cousin and the subterranean psychology of Maine de Biran. Biran had dedicated his writing to an embodied account of the self, which he developed in conversation with physiology and early brain studies. In resurrecting Biran as the source of "spiritualist realism or positivism," Ravaisson urged his peers to broach a more serious dialogue with the natural sciences so as to explore the corporeal facets of selfhood.[19] Jules Lachelier recognized that "the most sincere defenders of spiritualism in France do not hesitate to pay tribute today to Maine de Biran as their true master after Descartes."[20] In a letter to Ravaisson, Lachelier emphasized the division: "Aren't there in reality, since 1815, two philosophies: on the one hand, that of Mr. Cousin and his school, on the other, that of Maine de Biran. . . . Isn't it the silent struggle between these two philosophies, and the gradual triumph of the latter over the former, which you did not sufficiently indicate in your report?"[21]

My aim in this chapter is to trace the ruptured history of spiritualism. Its arc of was more sinuous than it was straightforward. Conflicting conceptions of the self propelled philosophical tendencies that came to a head in Ravaisson's report. These were encapsulated in two master concepts: Cousin's was *the rational sense of self*; for Biran, it was the *sensation of muscular effort*. Between them lay rival accounts—one substantial and introspective, the other kinetic and somatic—which respectively took root in the psychologies of Cousin and Biran.

This dispute endured in the aftermath of the report's release when a wave of figures embraced "spiritualist realism or positivism." They did so in large part as a reading community around Biran's resuscitated writings, which had gone neglected—and even suppressed—for the first half of the century. Formulated in the post-Revolutionary era, Biran's thought had to wait for a post-Cousinian future before it blossomed. Ravaisson, Lachelier, and Émile Boutroux rekindled Biran's fleshly psychology. They were not the lone representatives of the French philosophical establishment. After Cousin's death in 1867, others in the Parisian university system continued to wave the eclectic banner. In his 1877 treatise, François Magy defended the "relation that manifests . . . naturally between rationalism and spiritualism."[22] But Ravaisson, Lachelier, and Boutroux pursued a different path. These figures took from Biran the resources to locate the self's spiritual powers in the creativity of the natural world. They appropriated the life sciences to emphasize not determinist, but contingent principles of causality. Their intellectual trajectory was not confined to France. They also studied in Germany.[23] In dialogue with the inheritors of Kant's philosophy, these thinkers subjected French spiritualism to critical revision. They took Biran's philosophical anthropology as a point of departure and arrived at a spiritualized ontology of nature.

New movements claim new origins. The story of spirit was not one of chronological succession across the nineteenth century. In the pages that follow, I tell this story by following thinkers' understanding of their own historical pathways. These were by no means linear. "Spiritualist realism or positivism" developed out of the reimaginings that Ravaisson, Lachelier, and Boutroux brought to a shared philosophical archive, in which the spiritual and the material were entangled.

Maine de Biran on Motility, Selfhood, and Psychology

Philosophy was not Biran's initial pursuit. He was a noble who served as a guard in the court of Louis XVI from a young age. At the outbreak of the French Revolution, he defended Versailles during the October Days of 1789, when a mass of seven thousand women marched to the palace when their demands for bread were refused amid a grain shortage. Following the Revolution, Biran became active in government. Under the Directory, he represented the Southwest department of Dordogne in the lower house of parliament. Biran continued to serve the region through his life, first as a legislator after Napoléon's coup in 1799 and then as a member of the House of Deputies in the early years of the Bourbon Restoration in 1815.

Philosophy offered refuge from the tumult of politics. In Paris, Biran exchanged ideas in the salon of Madame Helvétius. In 1806, he organized a metaphysical society where diverse minds convened, including the naturalists Georges and Frédéric Cuvier; the philosophers Joseph-Marie Degérando and Pierre-Paul Royer-Collard; the physicist André-Marie Ampère; and the future prime minister François Guizot.[24] By 1816 the society included a young Cousin, whose "passionate mind," as Biran recounted, fueled "nothing less than the hope for a true philosophy among us."[25] Ideas of spirit were incubated in these intimate enclaves. Unlike Cousin, whose writings came to pervade French culture by the middle of the nineteenth century, Biran had little ambition for his work to circulate among a mass readership.

Biran's central idea was that one's sense of self-identity grows out of the muscular sensation of effort. His account of motor movement presumed that the experience of subjectivity derived from no origin other than its own impulsion.[26] Embodied volition involves auto-affection: the will's reflexive self-awareness. Biran articulated the idea in *The Influence of Habit on the Faculty of Thinking* (1802), a work he originally wrote for a competition organized by the Institut de France.[27] It won. Yet Biran published the book anonymously. "He had the taste for writing and the rather strange habit of beginning over again the same work without end," Paul Janet noted, "but he never had the taste for publishing, he feared it."[28] Biran argued that habit sustains a visceral feeling of interiority by imparting direction to the will. His view contrasted with a long-standing understanding of habit as blind repetition. Descartes and Kant, for instance, had condemned habit as mechanized routine. For Biran, automaticity lends a vital and even virtuous disposition to action. Underlying his conception of habit was an anthropological picture of the self as a continuous, lively, and even laborious exercise of effort modulating over time.

Biran developed his motor psychology in dialogue with empiricist doctrines. In *An Essay concerning Human Understanding* (1690), John Locke posited two sources of ideas. Sensations tell us about the external world. Reflection tells us about consciousness. For Locke, we acquire certain ideas only from sensation (namely: extension, shape, and motion); others we acquire from reflection (memory, judgment, knowledge, and faith); certain ideas depend on both sensation and reflection (pleasure, pain, being, power, unity, and succession). In France, Étienne Bonnot de Condillac advanced sensationalism, a radicalization of Locke's empiricism.[29] In his *Treatise on Sensations* (1754), Condillac claimed that sensation is the lone source of ideas. He did not countenance reflection as an independent power. Our reflective capacity, what Condillac called "attention," developed from affective states. As a thought experiment, he asked how a statue (an adaptation of Locke's tabula rasa) might gradually acquire knowledge. The statue's most basic sensation would be smell. Olfactory stimuli would furnish a primitive sense of pleasure and pain. The sense of touch would afford the statue knowledge of objects existing in space. For Condillac, we acquire self-consciousness, and come to situate ourselves in a physical milieu, thanks to the progressive accumulation of sense impressions. Subsequently, the French ideologues (as they called themselves) elaborated a science of ideas around of the study of linguistic signs. For example, in *Elements of Ideology* (1801) Antoine Destutt de Tracy categorized ideas into four faculties—sensation, memory, judgment, and the will. Like Biran, Tracy acknowledged that the will is unique, the "faculty we have of moving ourselves."[30] Although sympathetic to these variations of empiricism, Biran made a significant exception. To his mind, the inner sensation of effort was wholly unlike the sensations that derive from the external world.

The sensation of muscular effort is different. It is not punctual but processual. According to Biran, acting freely does not hinge on a yes or no choice; nor is it a binary on/off. Activating the will involves degrees of intensity, which modulate over time. Intensive gradations follow what Biran called the double law of habit: "the less we feel, the more we perceive."[31] In other words, repeating embodied practices desensitizes affection while improving intellection. Biran explored the law's effects on two kinds of habits. Passive habits diminish the degree of thought required for a task. I have learned to ride a bike when I no longer concentrate on coordinating my balance. The passive habit takes over as my feet ride the pedals instinctively. Active habits require willful exertion. Initially, I hear a new language as a cacophony. I commit myself to paying attention to each sound in order to master the vocabulary. I have learned the language when the sounds now strike my ears as a system of signs imbued with meaning. Biran had a monistic conception of active habits. They do not add content to thought; habits help us to discern the nuances already inherent in sensory experience.[32]

Active habits carry with them more than an affective psychology. Biran believed that they also open onto the metaphysics of selfhood. Embodied activity furnishes a primordial idea of interiority, which emerges out of the tension between effort and resistance. "Effort necessarily carries with it the perception of a relation between the being who moves or who wants to move, and an obstacle opposed to its movement. Without a subject or a will that determines the movement, without a term that resists, there is no effort, and without effort, no consciousness, no perception of any kind."[33] This dyadic (and dynamic) relationship was integral to Biran's idea of the self. Effort activates the enfleshed pleasure of directing my energy outward. In so doing, I come into corporeal contact with my finite limits. My body feels itself feeling. Together, effort and resistance are "two terms of the relation necessary to ground the first simple judgment of personality: *I am*."[34] Henri Gouhier clarifies how this phenomenology of effort was foundational to Biran's metaphysics of habit: "The basic example is not at all the effort exerted in order to raise a weight or to break a stick, that is to say the experience of a conflict between two forces, one internal and the other external. . . . Muscular effort gives me the feeling of a force deployed against an organic resistance, without any reference to an external object; their opposition is in no way that between subject and object, but between the active and the inert, the one and the multiple."[35]

For Biran, selfhood is riven, divided between the (active) self that wills and the (passive) self that resists. The conceptual distinction upended the empiricist understanding of sensation. Whereas passive sensations of the external world pertain to the impressions received via the body's individual organs, active sensations generate a holistic feeling of a unified self, of an "I who moves, or who *wants* to move."[36] Biran hewed to Kant's concept of the transcendental unity of apperception, according to which all experience is synthesized in a single self-consciousness. For Biran, however, unity is not an a priori condition of experience. He considered habit to be self-conditioning. The continuous passage between effort and resistance (as our habits wax and wane) generates conditions of experience immanent to experience itself.[37] Paradoxically, Biran derived the unity of selfhood from the duality at the heart of the will.

Biran also devised his motor psychology in conversation with early physiology. Xavier Bichat (1771–1802) and Pierre Jean-George Cabanis had drawn a distinction between the body's "inner" and "external" organs. The inner organs (e.g., the liver, stomach, lungs, colon, heart, and kidneys) maintain the organism through circulation and digestion. The external organs (e.g., the senses, nerves, larynx, and "voluntary muscles") furnish the sensory material for perceiving the outside world. Bichat and Cabanis parted with the prevailing anatomical paradigm that distinguished organs according to their physical structures. These medical thinkers instead bolstered a physiological paradigm predicated on organs' functions.[38] Cabanis was especially interested in "the systematic development of [man's] organs with the analogous development of his sentiments and his passions, relations from which it clearly results that physiology, analysis of ideas, and ethics are but the three branches of a single science, which may be justly called *the science of man*."[39] For Cabanis, the inner organs generate bodily affects that depend on age, sex, temperament, illness, diet, and climate. For instance, "in the bad quarters of cities, puberty is not given enough time to appear. It is hurried, and its effects are usually confused with the early habit of licentiousness."[40] By contrast, "within pious and strict families, where children's imaginations are directed toward religious ideas, the amorous melancholy of their puberty often appears to become mingled with ascetic melancholy."[41] Cabanis's point was that the inner organs coordinate corporeal experience in a manner completely unlike what he called the "distinct impressions received by the sense organs."[42] The latter constitute data of perception. Cabanis thought they were superficial aspects of the self; by contrast, the inner organs influence the body's whole condition. It was a small step for Biran to conceptualize muscular effort as a holistic mind-body activity made possible by the inner organs.[43]

Bichat's insights had an equally formative influence on Biran. Bichat traced the inner and external organs back to dual structures within the organism, what he respectively called "organic" and "animal" life. The structures were critical to histology, the branch of physiology that Bichat founded as the study of organ tissues. Organic life inheres in inner organs; these facilitate metabolic processes of assimilation and excretion.[44] In both vegetables and animals, organic life sustains the organism to "live within itself only."[45] Animal life is a function of the external organs, which bring the organism "as it were, out of itself: it is the inhabitant of the world, and not, like a vegetable, of the spot which gave it birth."[46] To Bichat's mind, it followed that each class of organs engenders distinct modes of sensory experience. Because external organs (e.g., the eyes, nose, ears, and hands) are symmetrical, comprising a right and a left half, they work in tandem to transmit harmonious sensations to the brain. A lesion to one eye, for example, affects the totality of the visual field. The internal organs are asymmetrical. Neither the stomach nor lungs depend on complementary organs to produce a unified sensation. Bichat's physiological schema was the foundation of a parsimonious sensory phenomenology. The external organs yield discrete sensations of the external world, while the inner organs generate fluid, interconnected sensations of the self's interiority.

The young sciences of the human organism offered Biran indispensable conceptual resources. From the physiological functions of inner and external organs, he refined the duality of affective experience. On the one hand, activity involves a sense of the will's seamless interplay with the body; on the other, passivity leaves us with scattered and haphazard sensory impressions. The dualism anchored Biran's ethics. Habits erode the initial resistance met by effort. But as we actively exercise our will, a secondary resistance ensues. Rote repetition takes over. "Its obstinacy is proportional to its blindness."[47] Error follows from unmotivated thought. Rules that at first require attention become formulae, leading us to overlook subtleties and singularities. For Biran, freedom is not just an intellectual affair (yes or no). It consists of a continuous and renewed effort to overcome habits, whose force is both the condition of freedom and its foremost enemy. His ethics set the will against itself.

Biran was by no means uncritical of science. From 1806 to 1810, he led the Medical Society of Bergerac.[48] There, Biran explored cranioscopy, the foremost field of brain research in the early nineteenth century. Friends gathered at the society to discuss the eccentric Viennese physician Franz Joseph Gall (1758–1828), who classified personality traits in the skull's protrusions.[49] Gall's indelible contribution to neurophysiology was the concept of cerebral localization, according to which specific areas of the brain are responsible for distinct mental functions.[50] The spark for his idea came in grade school, where Gall noticed that the smartest students, those able to recite material at length, had bulging oxen-like eyes. Subsequently, he identified the orbital cavities as the region indicating the powers of linguistic memory. Cranioscopy was not an experimental science. Gall palpated skulls and compared their size—observational methods that have since been discredited.[51] His charts of cranial structures featured maps of heads' contours, indicating the locales of characteristics such as benevolence, cupidity, self-esteem, and subservience. Such visual technologies provided eugenicists with a potent weapon to organize gender and racial hierarchies according to people's brain types. In France, Johann Spurzheim (1776–1832) facilitated the dissemination of the system that he named "phrenology." Biran took an interest when Gall and Spurzheim first presented their work on the nervous system and the brain at the Institut de France in 1808.[52]

What drew Biran's curiosity was neither the physiognomic diagrams nor the methods of skull measurement. At the medical society, Biran addressed Gall's neuroanatomy. Gall had upended the widely held view that the brain was the epicenter of the nervous system. According to his alternative framework, the brain constituted one of several neural crossroads. Ganglionic nerve endings formed bulbs (knots of neural tissue), which indicated prolonged nervous activity. They protruded most noticeably on the skull, making the bulbs apparent from the outside. Biran, however, believed the framework was incomplete. If the nervous system were decentralized, as Gall claimed, then the skull would exhibit only those nerves that terminated in the cortex. Drawing on Bichat's physiology of organic and animal life, Biran argued that nerves transmitting animal functions extended all the way to the brain. By contrast, organic functions traversed only the spinal cord. Biran identified two nervous structures where Gall entertained only one.

For Biran, the duality within neuroanatomy engendered two kinds of ideas (or better, two dimensions of spirit). The spinal nerves govern "affective ideas": the cranial nerves support "intellectual faculties."[53] The former express dynamic processes (namely, volition), whereas the latter represent discrete objects (such as entities in space). The distinction mapped respectively onto "those [ideas] which arise exclusively from the internal sense, and which apprehend nothing of the outside, and those which, on the contrary, have their reason in exterior observation."[54] This general distinction between internal and external dimensions of selfhood came to be a hallmark of spiritualism. For Biran, it responded to a particular problem, which was his diagnosis of Gall's approach to the self. The cranioscopic framework was in the grips of an *external* viewpoint, one that exclusively represented mental functions as if they were spatial entities that could be localized in separate cranial areas. Consequently, Gall had neglected the fluid nature of subjectivity. For Biran, the affective dynamism within oneself could be felt only via *internal* reflection. From this viewpoint, we experience our selfhood as a process, oscillating between passivity and activity—a tension occluded by cranioscopy. "If the affective faculties, like the intellectual faculties, have their seat in the brain, then where does this opposition and struggle come from, which we feel in ourselves between two principles of movements and determinations: that power of desire, the true motor force, at times dominant over that of passions and instincts; and appetites that pull in the opposite direction as in the stoic sage or at other times subjugated by those truly unhappy passions that seem to drag us away by a kind of *fatum*?"[55]

Biran's philosophical anthropology placed the dual structure of the will at the center of the self. Even though he articulated that duality using the idiom of "faculties," Biran struggled to relinquish the psychological doctrine. At the time, it was still believed that the mind was endowed with innate capacities to apprehend distinct kinds of knowledge.[56] Conceived by scholastic philosophers of the Middle Ages, faculty psychology partitioned the mind into separate powers: sensation, imagination, intelligence, and the will. Avowed faculty psychologists of the eighteenth century such as Thomas Reid (1710–96) and Christian von Wolff (1679–1754) posited additional powers, including testimony, sympathy, and intuition. Each was believed to constitute a distinct mental entity with clear boundaries. We find in the above passage, however, that Biran instead sought to construe the dual structure of the will as variations of *motion*. In his words, "principles of movement" determine our basic capacities. In fact, Biran argued that it was Gall who relied on facultative psychology, tethered as he was to a picture of mental traits situated in neatly separated compartments of the brain.

Articulated in dialogue with physiological sciences of the early nineteenth century, Biran's motor psychology laid claim to embodied ideas of spirit. It also prefigured a monumental turn away from faculty psychology and toward sensation and motion, categories that proved decisive in the rise of physiological psychology. That is a significant reason why, as we will see, the dynamic picture of spiritual powers—a dynamism that flowed from the body's motor exertion—came to animate spiritualist thinkers' engagements with the human sciences later in the nineteenth century.

Victor Cousin's Authority over Spirit

Biran's motor psychology became known only belatedly. After his death in 1824, four cartons of manuscripts fell into Cousin's hands. Joseph Lainé (1768–1835), president of the Chamber of Deputies under the Bourbon Restoration of Louis XVIII, was originally entrusted with the estate. But Cousin fervidly desired to publish the papers. An initial 1834 volume presented Biran's thought within the parameters of the eclectic doctrine.[57] These were, first, that the true activity of consciousness depends on the will; second, that the will is the self, "the personality and all the personality"; third, that the self is the first cause of all mental activity: "The person, the will, and causality are therefore identical."[58] Cousin's interpretation was not inaccurate. However, he overlooked Biran's engagements with his empiricist and medical interlocutors.

Critics alleged that Cousin had distorted Biran's legacy in order to safeguard his own eclectic philosophy. The socialist Pierre Leroux (1797–1871) decried Cousin for having "held for twelve years the case of manuscripts that had been entrusted to him, without making them known to the public; for only after two lustrums did he publish some of them. . . . But isn't twelve lost years, when the discoveries of a man like Maine de Biran are at stake, just baffling?"[59] The accusation was clear. Cousin had deliberately withheld the release of Biran's writings for fear that they might compromise the anti-naturalist bent of eclecticism. The historian Patrice Vermeren asks, "Was it not because its publication would have overshadowed [Cousin], reducing his own glory to have shattered the eighteenth century, while also posing the danger that one could have found in Maine de Biran the germ of another spiritualism, more rigorous and coherent than eclecticism?"[60] Paul Janet reflected, "It seems that the moment had already passed for the philosophical germ contained in these writings to be able to bear fruit."[61] As a result, readers remained largely ignorant of the full scope of Biran's motor psychology. For much of the century, ideas of spirit remained alloyed to Cousin's idiosyncratic vision of eclecticism.

A more complete four-volume publication was released in 1841.[62] It was read as Cousin's attempt to supersede Biran's embodied view of the self. "Now, in order to successfully bring together all the philosophy of [Biran's] psychology," Cousin wrote, "the first condition is for psychology itself to be completed, that it reproduce all the facts of consciousness."[63] These were the pure categories of the understanding (quantity, quality, relation, modality). Following Kant, Cousin believed the categories were necessary presuppositions of all knowledge. They constitute rational requirements of experience. Against Biran, Cousin argued that the will could not ground the categories of the understanding since volition is a subjective feature of the person.[64] Reason, by contrast, is not one's own. In the eyes of Cousin, Biran had failed to synthesize the personal exercise of the will with reason's universal authority.[65]

Synthesis was the goal of Cousin's eclecticism. He brought together what he took to be the three great modern philosophical systems of the seventeenth and eighteenth centuries: John Locke's empiricism, Thomas Reid's philosophy of common sense, and Kant's transcendental idealism. Together, they secured the metaphysical foundations of selfhood: the "triplicity" of sensation, will, and reason. By studying each of these psychological faculties in the context of its historical elaboration, students of eclectic spiritualism would arrive at a synthetic ontology, "voluntary activity leading to mankind, sensibility to nature, and reason to God."[66]

John Locke was significant for having introduced the notion of consciousness. The English word "consciousness" entered the French language as "con-science" thanks to Pierre Coste's 1714 French translation of Locke's *Essay*: "*Con-science* makes the *same person*."[67] Consciousness imparts a unitary form to experience so that it can be predicated on an identical self over time. For Cousin, Locke had laid the groundwork for the eclectic edifice, which took selfhood to be the enduring substance that undergirds the flux of sensations and thoughts.

The Scottish philosopher Thomas Reid had set in place the undeniable reality of the external world. In his *Inquiry into the Human Mind and the Principles of Common Sense* (1764), Reid claimed that rational discussion presupposes certain truths that are immune to skepticism. These truths include our direct contact with each other and with our common world. Instead of offering a logical defense of the external world's reality—and thus defeating the skeptical charge that had been a nagging problem of modern epistemology—Reid argued that the skeptic failed to participate in commonsense conversation. Cousin went further by claiming that knowledge of the self and its willed activity are equally indubitable. For Cousin, our personal sense (*sens intime*) discloses the experience of an inner self on which rational interaction relied.[68]

Kant's critical philosophy was the third system that Cousin believed "would introduce us to the depths of a problem which has escaped the other schools."[69] Cousin was first exposed to idealist philosophy in the salon of Madame de Staël (1766–1817), where writers met alongside aristocrats to discuss Kant, Jacobi, Herder, and Fichte.[70] De Staël had illuminated their philosophical pathway. "She opened up the doctrine so dear to Cousin by treating nearly all philosophers preceding [Kant] as materialists, some for having frankly espoused it, and others for having been committed to it without knowing."[71] Following his voyage to Leipzig in 1817, Cousin lectured on German philosophy in the 1820s at the Sorbonne.[72] He found in the transcendental unity of consciousness a principle to establish the identity of the self, a foundational truth anterior to sensory experience. Moreover, Cousin sought to enrich idealism by reconciling the conceptual divisions of the *Critiques*. Kant had divided theoretical and practical reason in order to leave open the possibility of faith. Reason could legislate experience only as it appears before the mind. Invoking Reid's commonsense philosophy, Cousin thought it absurd that we have contact only with phenomena. He also sought to penetrate the noumenal realm of God. "Metaphysics, for Cousin, could not continue for long as an abstract and logical science. Its eminently enthusiastic spirit would come to life and ignite itself in the face of the absolute."[73] Eclecticism would thus comprehend the rational kernel of Christianity.

The academic establishment embraced the eclectic edifice. Philosophers with posts in the University of Paris during the 1860s included Jules Barni, Francisque Bouillier, Paul Janet, and Léon Ollé-Laprune. They took the disembodied exercise of reason to be the sine qua non of selfhood. Moreover, they framed eclecticism as a historically complete enterprise. Among nineteenth-century thinkers, Cousin was hardly alone in his ambition to achieve philosophical systematicity. But it came at a cost. University philosophers expressed little interest in physiological advancements that might be brought to bear on the rational (and final) understanding of selfhood. Emblematic was the appointment of Elme-Marie Caro (1826–87) to a professorship at the University of Paris in 1864. At his public reception, the philosopher gave a lecture singing the praises of eclecticism. "Loyally accepting all the facts successively discovered by the positive sciences without altering their character, without limiting their true range, that is our strict duty."[74] Caro inveighed against materialism, which he described as "the philosophy of chance and fortune that takes hold of some still uncertain and incomplete facts. . . . That is our right as well."[75] The spiritual study of the self was thus metaphysically sealed off from the insights of the life and human sciences.

All the while, Biran's manuscripts underwent publication at a glacial pace during the middle of the nineteenth century. Many remained in the French Southwest town of Grateloup-Saint-Gayrand. In 1843, when writing his study *The Genius of Maine de Biran*, the Swiss philosopher François Naville (1784–1846) had received thirty-five pounds of manuscripts from

Biran's son, Félix de Biran. "Having been thrown away indiscriminately as paperwork in a waste bin, they were carried back to a grocer's home by one of his servants."[76] Naville abandoned the book and dedicated his efforts to publishing the carton. He made Biran's writings on psychology available in their near entirety in 1845.[77] Naville's death in 1846 led his son, Ernest Naville (1816–1909), to take over the project.[78] The Navilles' work was described as a "revelation" for having corrected Cousin's narrow presentation of Biran's thought.[79] Biran's engagements with physiology and cranioscopy, however, were published in their entirety only in 1887.[80] It was late. By then, experimental advancements in the brain sciences had surpassed Gall's bygone observational methods.

Nonetheless, Biran's insights offered a framework for subsequent thinkers to bring concepts of spirit into conversation with neurologists and physiological psychologists.[81] Biran did so, as Jerrold Seigel aptly notes, by introducing a positivist orientation into spiritualism, which "took the evidence of mental activity itself as a basic fact of psychology, thus challenging the materialist assumption that the primary phenomena of mental experience were sense-impressions, whose causes lay outside the subject."[82] Although underappreciated during his own lifetime, Biran's motor psychology supplied an antidote to what many later came to see as Cousin's conservative power over the French academic establishment. The dissident professor Joseph Ferrari (1811–76) lambasted, "Mr. Cousin was born in the university; we don't know him to be a part of any other family, and he carries his condition like a monk."[83] Biran's belated publications followed circuitous routes that ultimately cascaded into a new spiritualist current in the second half of the nineteenth century.

From the Psychology of Selfhood to the Ontology of Nature

The prospect of spiritualists' departure from eclecticism had emerged in the early years of Cousin's regime. In 1834, Cousin sponsored Ravaisson, whose first essay on Aristotle's *Metaphysics* won a competition hosted by the Académie des sciences morales et politiques. With Cousin's blessing, his young student traveled to Munich, where he attended Friedrich Schelling's lectures on aesthetics. Their influence on Ravaisson was ambiguous.[84] Both saw philosophy as an artistic practice that traversed the realms of nature and freedom.[85] The greater influence on Ravaisson's thought was certainly Biran, whose embodied psychology sowed the seeds of "spiritualist realism or positivism."

Ravaisson believed that eclecticism had sullied the spiritual powers of selfhood. In an 1840 article, he argued, "Mr. Cousin declares that he adopts

Maine de Biran's doctrine of the identity of the self, or of the identity of the personality and the will. . . . But at the same time, it seems to us that he has denatured and annulled it on account of the restrictions that he imposes."[86] Ravaisson interpreted Biran's concept of muscular effort—*pace* Cousin—not as a personal feeling. "Effort supposes . . . an anterior tendency which, by its auto-development, provokes resistance; it is the original activity, prior to effort, which, reflected by resistance, comes to possess itself by itself in a voluntary action."[87] Nature, not reason, was the universal ground of selfhood. Ravaisson read Biran through a vitalist lens in order to escape a vicious circle into which his psychology risked plummeting: whence the original source of effort? What is the first mover behind human motility? Ravaisson claimed that habit channeled a single energy, a creative power flowing from nature to the self. Jules Lachelier went on to remark, "I don't need to remind you of the article . . . in which Ravaisson connected his writing to the philosophy of Maine de Biran. There you have our point of departure."[88] By setting Biran's anthropology of the self in an ontology of nature, Ravaisson bridged the disjointed spiritualist heritage, from the Biranian past to a post-Cousinian future.

Ravaisson's opus was *Of Habit* (1838), the doctoral thesis he submitted to Cousin. The central idea was that habit involves not only a dynamic fluctuation between effort and resistance; it also depends on the fecundity of nature. Ravaisson described habit as "a *natured* nature, the product and successive revelation of *naturing* nature."[89] Effort channels nature's creative impetus while resistance bears the weight of nature's gravity. He lent nuance to Biran's dualism between passivity and activity. For Ravaisson, the process of habituation constitutes "a graduated continuum from the lowest levels of activity in more passive, sensory experiences to its highest levels in the clearest consciousness and voluntary apperception."[90]

The continuum was illustrative of Ravaisson's hierarchical ontology. In *Of Habit*'s first half, he traced the ascent of nature from brute matter through vegetable, animal, and ultimately human life. His aim was to show that habits pervade the organic realm. The freer the organism, the greater its potential for habituation. Ravaisson noted that even bamboo shoots habitually mold to the pole around which they grow. In animals, habits impart freedom in the form of spontaneity—the interval between an organism's reactions to the environment and the actions it carries out on its own volition. He drew on the contrast between spontaneity and repetition that was central to the work of vitalist thinkers, including the seventeenth-century animist Jan Baptist von Helmont; the eighteenth-century physiologist Paul Joseph Barthez; and Xavier Bichat. They shared a commitment to the existence of a vital principle that explains the variability of living organisms in contrast to the mechanical laws of inorganic matter.[91] Ravaisson claimed that the double law of habit had its source in organic spontaneity. "The change that has come . . . from the outside becomes more and more foreign to it; the change that it has brought upon itself becomes more and more proper to it. Receptivity diminishes and spontaneity increases."[92] In humans, spontaneity is pushed to the limit. The will activates an "internal potentiality" surpassing the body's recurring nutritional functions.[93]

In the second half of the book, Ravaisson traced the descent "from the clearest regions of consciousness [as] habit carries with it light from those regions into the depths and dark night of nature."[94] Drawing again on Biran's motor psychology, Ravaisson claimed that awareness of selfhood depends on the sensation of effort. Effort enlivens thought. When habit diminishes the effort required to execute actions, thinking fades into brute repetition. Ravaisson's picture of thought was anti-representational. "Ideas are no longer representative, it is the idea in action, the actualization of the ideal in this confusion of the end and of movement that is its tendency."[95] For Ravaisson, consciousness is a protean motility; it does not exist "in the head" but is, rather, enveloped by a creative energy that precedes and exceeds the boundaries of the body. "Habit is thus, so to speak, the infinitesimal *differential*, or, the dynamic *flexion* from will to nature."[96] Humans are creatures that can guide their habits and can, therefore, become intimately aware of the interplay between nature and selfhood. Pierre Montebello describes this as Ravaisson's anthropomorphism: "What is in man is in all things, not because it is in man, but because he is in all things. This superior anthropomorphism seizes man at his root: and his root is the cosmos."[97] Ravaisson thereby set in motion an enduring ontological conviction of spiritualist thought: that humans are nature's zenith, the beings who bring the creativity inherent in biology to its highest state of reflection.

Ravaisson's final aim in *Of Habit* was to articulate the aesthetic perfection of the will. He suggested that the authentic exercise of freedom depends on the refinement of embodied habits. In willing, one ought to exert enough effort so as not to overwhelm resistance. In perhaps the most eloquent passage of the book, Ravaisson described this delicate equilibrium:

> As resistance fades, there is no longer anything to reflect the principle of action back onto itself; nothing calls it back to itself. Its will is lost in the excess of its freedom. The subject experiencing pure passion is completely within himself, and by this very fact cannot yet distinguish and know himself. In pure action, he is completely outside of himself, and no longer knows himself. Personality perishes to the same degree in extreme subjectivity and in extreme objectivity, by passion in the one case and by action

> in the other. It is in the intermediate region of touch, within this mysterious middle ground of effort, that there is to be found, with reflection, the clearest and most assured consciousness of personality.[98]

The "mysterious middle ground of effort" amounted to what Ravaisson called "the law of grace."[99] Grace brings nature and the will into agreement. For instance, a ballerina's dance expresses a seemingly effortless levitation.[100] The delicate continuity between effort and ease was an aesthetic ideal that Ravaisson also conveyed in theological terms: "Nature is prevenient grace. It is God within us, God hidden solely by being so far within as in this intimate source of ourselves, to whose depths we do not descend."[101] Although it issues from within, grace manifests its divine splendor on the visible body. Herein lay the aesthetic imperative of "spiritualist realism or positivism": nature inscribed on the flesh.

Ravaisson was no stranger to art. In 1849, he became a member of the Académie des inscriptions et belles-lettres (the official philosophy institute, the Académie des sciences morales et politiques, twice rejected his candidacy). From there, Ravaisson assumed a government post charged with reforming drawing courses in schools. The Ministry of Public Instruction saw the courses as a means to ensure the hegemony of "good taste" in France (with an eye to surpassing the cultural prowess of the nation's rival across the English Channel). Ravaisson put in place a graphic arts curriculum that inculcated *intuitive* techniques. Unlike the analytic method of geometric repetition, whereby students would trace the lines of exemplary images, Ravaisson promoted visual discernment. "Visible appearances constitute a silent language for which our eye is the organ and by which the spirit producing and manifesting itself in forms makes itself known to our spirit."[102] Students were taught to pierce beneath the surface of drawings to ascertain their inward movement. The pedagogical task was mimetic. Students would study textbooks with images of works by past masters.[103] Ravaisson's method (alluding to Leonardo da Vinci) was to intuit the *serpeggiamento*: the line that served as a governing axis from which each brush stroke followed on paper. School lessons in drawing aimed at extracting the fluid continuity within mechanical representation. Ravaisson thus brought students' attention to the kinetic forms of beauty in classical works, in which "the character of movement and forms is grace rather than force."[104] Ravaisson's drawing pedagogy was of a piece with his metaphysical emphasis on the creativity inhering in nature. For this reason Bergson later suggested, "The whole philosophy of Ravaisson springs from the idea that art is a figured metaphysics, that metaphysics is a reflection of art."[105]

Although his work garnered wide recognition, Ravaisson never became a university professor. That such a brilliant philosopher spent his career as a state functionary has led some to speculate that Cousin barred Ravaisson from entering academia.[106] He took a position as the secretary of the Ministry of Public Instruction in 1838 under Narcisse-Achille de Salvandy, an outspoken opponent of Cousin. In 1840 Ravaisson was appointed inspector of libraries. From 1859, he served as inspector general of higher education. Ravaisson sat on the jury of the *agrégation* in philosophy.[107] The state-administered competitive examination accredited philosophy students, allowing them to pass on to doctoral studies. It had been used to vet an educational elite since its inception under Louis XV in 1766. Sitting on the jury, therefore, endowed Ravaisson with incredible power to mold young philosophers. "It was Ravaisson who taught us," Jules Lachelier wrote in a letter, "to conceive being, not by the *objective* forms of substances or of phenomena, but by the *subjective* form of spiritual action; and furthermore, this action ultimately lies in thought or will [*volonté*]."[108] Herein lay the kernel of "spiritualist realism or positivism," a movement that reimagined the spiritualist past and oriented it toward a new future. Ravaisson brought together a community of readers around Biran's work, showing that effort—central as it was to Biran's motor psychology—emerged from and participated in the natural world.

Jules Lachelier and the Aesthetic Order of Nature

Spirit and nature were central ideas in Jules Lachelier's writings. He highlighted the limits of mechanics, demonstrating that its causal models were not sufficient either to explain the natural world or to extirpate the immaterial dimension therein. In his opus *On the Foundation of Induction* (1871), he forcefully claimed, "The true philosophy of nature is a *spiritualist realism*, in the eyes of which every being is a force, and every force a thought which tends to a more and more complete consciousness of itself."[109] Lachelier thereby elaborated on Ravaisson's chief insight that freedom inheres in the inventive strata of nature.

Lachelier entered the École normale supérieure in 1861. Although Ravaisson was not a professor there, his influence was noticeable. The two would retreat from the Parisian academic scene to go on hikes in the forest of Fontainebleau. Their conversations fueled Lachelier's philosophical interests. For his thesis, he took up Biran's psychology. "I think I agree with everyone in considering him as one of the strongest supports of spiritualism."[110] The issue at stake was the problem of causality. Biran had claimed that effort furnishes its epistemological foundation. In the act of willing, the self feels its power as a cause and the embodied movement that ensues as an effect. The immediate apprehension of this copresence of cause and effect constitutes the primitive feeling of a necessary determination.[111] Lachelier poked holes in Biran's account. If the foundation of causality is the will, then could science explain the causal relations among natural phenomena? Lachelier answered in the negative. "It is difficult to abstract the absolute and necessary notion of substance from a relative and contingent fact of individual existence. There you have the vague character of a notion meant to suit both consciousness and exterior bodies."[112] Lachelier completed twenty-nine pages of the thesis before abandoning the project.

Yet, Lachelier did not abandon Biran. In the middle of his studies, the Ministry of Public Instruction reopened the philosophy agrégation after it had been suppressed under the Second Empire. Lachelier took the examination in 1863 and finished first in the new class of *agrégés*.[113] His success granted him a teaching post at the École normale. There, his lectures shaped a generation of students who came of age in the post-Cousinian intellectual climate. He gained a reputation as the vanguard of Kantianism. As Lachelier quipped to Ravaisson in a letter, he self-consciously "*Kanticized* all year long."[114] In fact, Lachelier was known for meeting students in his office, where a copy of Kant's first *Critique* rested on his desk, always open to the page stating that the "I think" accompanies all our perceptions.[115]

Lachelier's understanding of the self, however, relied on an idiosyncratic—to wit, spiritualized—reading of Kant. His approach developed, again, out of the problem of causality, which he tackled in *On the Foundation of Induction*. Lachelier began with efficient causes. These concern motion and change. For example, Isaac Newton posited gravity as an efficient cause. The law of universal gravitation held that objects attract each other according to a force that is inversely proportional to the square of the distance between them. Newton formulated the law by means of induction. He inferred a general law from his observations of particular falling bodies. But inductive reasoning is problematic. How do individual instances ensure the veracity of a universal law? Lachelier framed the problem according to Kant's conceptual requirements. Kant had posited in the first *Critique*, "the concept of cause . . . requires that something A be of such a kind that something else B follows from it necessarily and in accordance with an absolutely universal rule."[116] Lachelier sought to comprehend how necessary and universal laws might be grounded on the basis of contingent facts.

The natural sciences do not rely on efficient causes alone. Lachelier was more interested in final causes, which explain the spontaneity of living beings—that is, the relations between a whole and its parts. In biology, individual organs' functions are determined by their role in a physiological system. Lachelier drew on the French physiologist Claude Bernard (1813–

78), for whom, "In every living germ is a creative idea which develops and exhibits itself through organization."[117] For example, the "creative idea" of the circulatory system is to transport nutrients. Unlike the efficient causes of objects in motion, final causes determine the purpose of processes such as respiration and digestion. Crucially, Lachelier did not take final causes to constitute a *necessary* determination. Finality is "a flexible and contingent law in each of its applications: it requires a kind of harmony in the ensemble of phenomena, but it guarantees neither that this harmony will always be composed of the same elements, nor even that it won't be upset by any disorder."[118] Beyond efficient causes, final causes determine reciprocal variation in biological processes such as reproduction and retardation or evolution and devolution.

At stake for Lachelier was how these two kinds of causality hang together. Following Kant's transcendental method, Lachelier demonstrated that efficient causality and final causality are not empirical principles found in the world. They are conceptual principles of science. Their genesis owes to the work of reason. The significance of his approach turned on its divergence from British empiricism as well as Cousin's spiritualism. On the one hand, the empiricist school held that an associative bond forms in the mind when two events are perceived together regularly. In *A System of Logic* (1843), John Stuart Mill (1806–73) had argued that these bonds constitute the psychological foundations of efficient causality. Such was the basic premise of associationism. For Lachelier, however, inductive reasoning was not veracious if it depended only on factual regularity. Returning to Kant, "the effect does not merely come along with the cause, but is posited through it and follows from it."[119] Lachelier reasoned that a *foundation* must determine relations between cause and effect. On the other hand, Cousin had conceived final causality as an intellectual intuition. To conceptualize an organism's internal organization, the scientist would invoke a causal principle independent of his or her observations of the organism. Lachelier found this account of final causality to be too vague. It took the concepts of nature to emerge sui generis. He concluded, "Empiricism vainly endeavors to set a principle on the solid, yet narrow ground of phenomena; the opposite doctrine [eclecticism], in order to give this principle a wider base, builds in a void and only manages to state a need of consciousness while claiming to satisfy it."[120]

For Lachelier, induction relies on the exercise of reason. He did not take reason to be an abstract exercise that, following Kant, synthesizes particular facts of experience with the understanding's category of causality. Lachelier showed that reasoning is also a vigorous activity, one that is animated by the creative energies of nature. As one commentator noted, "The problem posed in *On the Foundation of Induction* is not that of induction, but that of the choice of this subject in order to demonstrate a philosophy of nature."[121] Lachelier claimed that reason does not operate outside of nature. The purposiveness apparent in organisms is not a purely ideal construction, what Kant characterized as a regulative (and not constitutive) use of judgment in the third *Critique*. Purposes are also evident in natural processes themselves; their "necessary determination is without a doubt something distinct from us, for it imposes itself on us and resists all the vagaries of our imagination."[122] Lachelier's spiritualist interpretation of Kant demonstrated that although the categories of reason issued from the spontaneous activity of the transcendental self, that very spontaneity issued from the creative forces within nature.

As Lachelier saw it, induction ultimately depends on final causes, expressive as they are of nature's inner creativity. By contrast, efficient causes reflect the superficial strata of natural phenomena. "The realm of final causes, by penetrating the realm of efficient causes without destroying it, exchanges everywhere force for inertia, life for death, freedom for fatality."[123] Although Lachelier articulated his account of induction in a transcendental vocabulary, he went further by arguing that contingency—and not only necessity—is a conceptual requirement of the sciences. Contingency constitutes the ground on which efficient and final causes cohere. For example, efficient causes determine the trajectory of a falling tree. But the tree's development, from the roots plunging into the soil to the growth of its limbs, depends on final causes. Final causality determines the intensive movement hidden by any extensive measurement of the organism. The former reflects the aesthetic order of nature. "Nature thus possesses two existences founded on two laws that thought imposes on phenomena: an abstract existence, identical to the science for which it is an object, and a concrete existence, identical to what one could call the aesthetic function of thought, which rests on the contingent law of final causes."[124] It is hard to underestimate just how radical Lachelier's claims were. In the nineteenth century, nature had been so thoroughly mechanized that it challenged the bounds of conceivability to imagine pliable and open-ended—in short, contingent—natural processes.[125] For Lachelier, the interconnection between spirit and nature was contingent. Their aesthetic harmony could be ascertained only from our human form of life. We are beings who elevate the contingency evident in nature to a constitutive principle of scientific inquiry.

Lachelier's admiration for the sciences did not always shine through his writings. Some criticized him for advocating the "bankruptcy of science." "At the École normale they teach the relativity of science and the supremacy of morals," one critic pilloried; "you find there, since Lachelier, all the young generations of students learning it while beginning their studies."[126] However, Lachelier hardly aimed to undermine scientific inquiry. He recognized the usefulness of mechanistic determination. In a letter to Ravaisson, Lachelier clarified, "Determinism is certainly not the truest system, and I even take it as absolutely false the moment it becomes exclusive, but on the other hand, it is the clearest, or rather, the lone clarity for our mortal eyes."[127] Lachelier considered mechanics to be a pragmatic tool; but he reached beyond its utility in order to articulate a more capacious ontology of nature.

On the Foundation of Induction endured as Lachelier's most celebrated book until his death in 1913. Despite its mass appeal, he insisted it was only a modest piece of philosophy. When the philosopher Gabriel Séailles (1852–1922) offered to write his biography, Lachelier declined. "I had, I believe, some philosophical ideas . . . hardly original, nearly all drawn from Descartes, Leibniz, and Plato as well as Aristotle, who were all, no doubt, more valuable; but these ideas were never strongly enough connected, nor even largely enough developed, to make them into a system, or even simply a doctrine."[128] Perhaps Lachelier did not espouse a system. But his thought was critical to the advancement of "spiritualist realism or positivism." First, he upended the Kantian conception of nature. Universality and necessity were no longer the foundation, but the surface of science. It is "rather universal contingency that is the true definition of existence, the spirit of nature and last word of thought."[129] Second, Lachelier took stock of the lacunae within the sciences' working categories and treated them as a springboard from which to vault the metaphysics of spirit. In so doing, he advanced a revolutionary understanding of the dynamic liaison between the vital processes of nature and the creative powers of selfhood.

Émile Boutroux and the Necessity of Contingency

Among Lachelier's most accomplished students was Émile Boutroux, a Catholic whose faith informed his commitment to humans' freedom within the natural world.[130] His philosophical orientation was clear at an early age. "As soon as I was at the École normale [in 1865], I wondered why philosophy had been confined to the section of letters, while all the great thinkers . . . had participated in the sciences"; he believed that if philosophy "is a matter of truth, and not only of sentiment, then it could not consider the acquisitions of the positive sciences as its strangers."[131] Boutroux radicalized the concept of contingency, showing how it was not exceptional to the mechanistic worldview, but immanent to the laws of nature.

Boutroux's early fascination with scientific experimentation was evident in his rich correspondence with Lachelier. "Take the example of water boiling. How does it boil? The bubbles of vapor form and leave the vase that is in contact with the fire; the bubbles burst for a certain time before arriving at the surface of the liquid."[132] Boutroux believed that experimental methods uncover more than just how water boils. "Science, Claude Bernard says, consists of *predicting* the phenomena of nature and *mastering* them. In other words, after having responded to the question *how*, you must respond to the question *why*? It is when we know the phenomena by their determinate cause that we can predict and master them."[133] Boutroux's interest in the structure of casualty resonated with his professor's work. But Lachelier encouraged Boutroux to keep going: "the question *why* might be susceptible to several meanings, and these meanings could be the object of metaphysics."[134] After Boutroux passed the agrégation in philosophy in 1868, Lachelier suggested, "you will be a dangerous professor, so shouldn't we send you on a mission."[135] Boutroux left to study under Eduard Zeller (1814–1908), the neo-Kantian professor at the University of Heidelberg. Upon his return, Boutroux waxed effusively about the proximity of the arts and sciences in German education—a model that he went on to promote in France.[136]

In his doctoral thesis, *The Contingency of the Laws of Nature* (1874), Boutroux built on Ravaisson's hierarchical ontology of nature.[137] Boutroux organized its strata according to their complexity: from inorganic matter to organic creatures, from animal life to human being, and from sensibility to intelligence. He claimed that the science corresponding to each stratum manifests greater contingency: from physics and chemistry, to biology, psychology, and finally philosophy. For example, biology makes use of physical and chemical principles but its constitutive complexities cannot be reduced to physics or chemistry. The hierarchy also hewed to Auguste Comte's ladder of knowledge.[138] Ascending the ladder rendered knowledge progressively more concrete and pertinent to a science of man. Descending the ladder, each science became more abstract, supplying the laws on which higher knowledge depends. Positivists' vision had garnered widespread support not only in France, but also across England, India, and the Americas, where the Brazilian flag still bears the motto "Order and Progress."[139] It was believed that scientific progress would lead humanity to relinquish religion, including its otherworldly residue in the form of metaphysics. Although he adhered to positivists' conceptual architecture, Boutroux eschewed universal principles. He reasoned that the sciences could not be totalized within the same system: "The universe is not made up of elements equal to one another, susceptible to being transformed into one another, like algebraic quantities. It is made up of forms bound together by gradations, i.e., additions that are altogether imperceptible."[140] Boutroux considered necessity and universality to be partial aspects of scientific reasoning; they do not hold absolutely. Like Comte, Boutroux positioned philosophy atop the hierarchy. Far from viewing it as the synthesis of scientific knowledge, however, Boutroux conceived philosophy as a creative vehicle to leap beyond the bounds of science.

At stake for Boutroux was the place of contingency within the hierarchy of sciences. At one end, mathematics reigns in contingency; its rules absolutely determine numbers. On the other end, the human sciences revel in contingency, since personal creativity outstrips the laws of matter and motion. Throughout, contingency is a conceptual requirement of all scientific reasoning. In order to distinguish a cause from its effect, a creative difference must obtain. "How can we imagine that the cause or immediate condition really contains all that is needed to explain the effect?"[141] Boutroux parted with Kant's conception of causality. He argued that contingency ensures the nonidentity of cause and effect. If a cause exhaustively explained its effect, then the cause would "never contain that wherein the effect is distinct from itself, that appearance of a new element which is the indispensable condition of a relation of causality."[142] The life and human sciences thus presuppose that a heterogeneous relation holds between a cause and its effect. In a Kantian vein, Boutroux justified the claim both by fact and by right. By fact (*quid facti*), scientists observe effects in nature that surpass their cause. "The seed that falls from the beak of a bird onto a snow-clad mountain may occasion an avalanche which will submerge the valleys below."[143] By right (*quid juris*), the sciences require contingency because the very concept of causality presumes that the conditioned engenders novelty in excess of its conditions.

Emblematic was the law of the conservation of energy. It held that the total energy of an isolated system remains constant over time. Developed during the nineteenth century by the physicists Robert Mayer, James Joule, and Hermann von Helmholtz, the law subsumed singular exchanges of energy under a necessary relationship. For Boutroux, contingency was a constitutive aspect. Energy would remain constant only when the quantitative relations of motion were abstracted from qualitative transformations. For example, water's transformation into vapor is made calculable by construing the process in the numerical terms of the amount of energy dissipated. A difference in kind is translated into a difference in degree. "The law of the conservation of force presupposes a change it does not explain, which it would even make unintelligible were it regarded as possessing undivided sway over primordial modes of matter."[144] The heterogeneity between cause and effect went suppressed in order to facilitate the calculation of aggregate energy. To Boutroux's mind, the law of conservation was overly abstract. It exemplified what he called the "static sciences." By contrast, "dynamic" science would employ a historical method to follow the microscopic changes in natural events. For Boutroux, history entailed the meticulous description of singular qualities. "Thus, every fact depends not only on the principle of conservation, but also, and in the first instance, on a principle of creation."[145] A truly dynamic science would ascertain the genesis of phenomena, tracing the concatenation of contingencies that give rise to lawlike regularities.

In the book's conclusion, Boutroux directed his account of contingency squarely against the Kantian conception of science. In place of what he called the "doctrine of reconciliation," which rescues human freedom by excluding it from the causal order of nature, the doctrine of contingency "does more than open onto freedom apart from the world, a field that is infinite, though void of objects which it can contact. It shatters the postulate which makes inconceivable the intervention of freedom in the field of phenomena, the maxim which states that nothing is ever lost and nothing created."[146] Boutroux's aim was to overcome the chasm separating the sensible realm of phenomena and the suprasensible realm of freedom. Since his early studies, Boutroux thought the division was untenable. An unpublished essay of his read, "From the moment that laws or properties have reality only in spirit, what exists beyond spirit would only be an existence, devoid of all species and manner of being; but such an existence is indistinguishable from nothing."[147] Boutroux saw our powers of scientific reasoning as the culmination of the contingency inhering in nature. Humans endow authority to scientific laws because we are a species that participates in the very natural processes of which those laws are a piece. In his words, contingency "lends itself to the conception of a freedom coming down from suprasensible regions to mingle with phenomena and direct them along unforeseen paths."[148]

The Contingency of the Laws of Nature marked a significant contribution to "spiritualist realism or positivism." First, Boutroux advanced a concept of the self and its reasoning powers that was rooted in a spiritualized ontology of nature. Second, he opposed this spiritualist idea of selfhood to the Kantian transcendental self, whose freedom was secured at the limit of the natural sciences. "In order to content our logical understanding, we let go of the spiritual proper, the real being: this lies, beyond its determinations, in the power to determine, in the creative faculty, in energy"; Boutroux claimed that this idea, "dear to Ravaisson . . . was one that Lachelier made into his own conviction."[149] The idea of contingency provided these thinkers conceptual scaffolding on which to build a bridge between the creativity of nature and the self's power of thinking.

"Spiritualist realism or positivism" brought about a rupture in the French intellectual scene. In response to the eclecticism of Cousin, whose vision of selfhood had unmoored spirit from the insights of scientific pioneers, a wave of thinkers reimagined their philosophical horizon extending forward and backward in time. In this context, Biran's embodied psychology became a post factum origin. Unearthed from beneath the eclectic regime, his writings opened a metaphysical wellspring that came to animate the work of Ravaisson, Lachelier, and Boutroux. These thinkers sketched new ideas of spirit that coalesced around a fecund ontology of nature. Their readings of Biran were not always faithful. They creatively appropriated his idea of motility, weaving it into a higher anthropocentrism—the apogee of a material world whose creativity was interwoven with the natural sciences' causal laws. In being drawn on, Biran's motor psychology made possible interpretations that testified not so much to its distortion, but to its durability—even urgency—in the historical circumstances of the mid-nineteenth century. In the eyes of Lachelier, this process of intellectual transmission followed divergent currents of spiritualism. "There is a kind of spiritualism, which, at face value, consists of simply placing spirit above nature without establishing the relation between one and the other. But there is a deeper and fuller spiritualism, which consists of searching within spirit for the explanation of nature itself, on the evidence of the unconscious thought that works within nature and becomes conscious within us, and that only works in order to bring about an organism through which it passes from . . . an unconscious form to a conscious form."[150]

The first spiritualism—that of the eclectic school—was a point of resistance for the thinkers who rallied around Ravaisson's *Report on Philosophy in France during the Nineteenth Century*. Yet, Cousin was not just a foil. Ravaisson, Lachelier, and Boutroux also elaborated on "spiritualist realism or positivism" in conversation with Kant's epistemology of science. Idealist philosophy was a formidable body of thought without which these thinkers could not have so forcefully enlarged their understanding of nature. In so doing, they brought about a new stage in the entwined meanings of spirit, selfhood, and science. This—the second spiritualism—reimagined its roots in an inherited French tradition. The concluding lines of Ravaisson's 1867 report averred: "If the genius of France has not changed, there will be nothing more natural for her than the triumph of the high doctrine . . . that freedom is thus the last word of things, and that, beneath the disorder and antagonisms which trouble the surface where phenomena occur, in the essential and eternal truth, everything is grace, love, and harmony."[151]

✵ CHAPTER 2 ✵

Measuring the Machinery of the Brain

To make the mind measurable; to translate thinking, feeling, and willing into countable units; in short, to quantify the self: these ambitions came to the fore of the French intellectual community in the second half of the nineteenth century. The authority of numbers reigned. Quantification promised to endow psychology—still a nascent field—with the precision, verifiability, and objectivity long believed to be reserved for physics, which had subsumed the material world under mathematics. If the brain's machinery could be measured, could numbers also apply to the self's spiritual powers? New experimental models aspired to do just that: physiological psychology clocked the speed of nervous transmission; psychophysics calculated the intensity of sensations; mental chronometry and psychometrics measured the duration of recognition, memory, and attention. Widespread enthusiasm for quantification led many to interpret the self's interiority as a clockwork machine. Indeed, for Théodule Ribot (1839–1916), the founder of experimental psychology in France, "the fact of consciousness has, like all other phenomena, a precise, variable, and measurable duration."[1] The quantitative imperative strived to make real what could not yet be taken for granted: that there are mental data, and those data could be calculated, accumulated, and organized into a body of psychological knowledge.

The ascendance of quantitative models lent traction to the materialist turn in French ideas of spirit. Translating spiritual activities into numerical terms brought about a conceptual dilemma. From the outside, the duration, intensity, and frequency of mental processes could be calculated. From the inside, there appeared to be a variety of nonnumerical data. These were *immediate* data of consciousness. The fluidity of experience disclosed facts that were all the more authoritative, so the thinkers in this chapter argued, because those facts were enmeshed with the intimate feeling of selfhood. For Jules Tannery (1848–1910), "Sensation is a phenomenon occurring in us, which we seize in ourselves by its interior side, as it were, and which rebels against any kind of measurement."[2] The mathematician's words were influential for Émile Boutroux, who argued that inner states eluded quantification. "We look in vain for psychic units susceptible to addition or subtraction like mathematical units."[3] He argued that there were instead *qualitative* data of consciousness. The feeling of volitional activity is experienced as a continuum over time; it exhibits a heterogeneous structure unlike homogeneous numbers. For Henri Bergson, inner experience flowed in the form of a *lived duration*, the commanding idea in his *Essay on the Immediate Data of Consciousness* (1889).[4] As the title suggests, the book argued that quantitative facts were only one variety of psychological data. Bergson embarked on the project after he gave up on a mathematics degree.[5] "I realized that mathematics could not explain time, as one perceives it within oneself; and so, everything that I had neglected up until that day as secondary became essential for me."[6] In a similar vein, Alfred Fouillée reasoned that science made mental phenomena "expressible by means of three fundamental quantities—length, time, and mass—without discussing their nature."[7] It was up to philosophers, Fouillée inveighed, to articulate the underlying qualities that made data intelligible. Despite their diverse vocabularies, these figures cohered around overlapping interests in the science of quantification. Their contributions to ideas of spirit followed from a sustained dialogue with psychology's mathematical models, which offered a *point d'appui* for them to articulate the qualitative data of selfhood.

Chronometry, psychometrics, and psychophysics prompted a radicalization of the "spiritualist realism or positivism" whose formation I traced in the last chapter. Among its tenets were: first, that there is continuity between natural and spiritual activity; second, that spirit depends on the will, the locus of the self. The figures in this chapter went further. They spearheaded a materialist moment within spiritualism by endowing the self's interiority with factual authority. For these figures, the affective and conceptual texture of volition exhibited an integrity that made effort, agency, and creativity into the source of a kind of psychological information distinct from yet also compatible with quantitative data.

It might be tempting to read the figures in this chapter as recalcitrant critics of mathematics who clung to traces of Christianity. On this reading, spiritualists guarded the unity of the self. Descartes had thought that matter could be divided into discrete units since its essence was extension. But he had vanquished numbers from the unextended soul, which had been created by a single act of God's grace. Lachelier appeared to defend the Cartesian legacy in his screed *Psychology and Metaphysics* (1885), when he criticized mathematical models: "The interior life of man has been reduced once again to sensation, having become the simple consciousness of an organic state."[8] Many interpreted the piece as a call to arms. I want to suggest, however, that spiritualists in the late nineteenth century were actually quite fascinated by the transformation of mental interiority into a matrix of empirical facts. Introspection had bequeathed the concept of an inner realm. The pioneering German psychologist Wilhelm Wundt (1832–1920) described his work as a "science of experience" (*Erfahrungswissenschaft*).[9] The search for a numerical ground of psychology made it more pressing than ever to spell out the architecture of mental interiority. Taking stock of spiritualists' investment in this problem casts these thinkers' constructive (and not only critical) motivations into stark relief: they did not simply spar with quantification but advanced a positive project to craft "spiritual" data predicated on a form of validity beyond that of quantity. On this reading, Lachelier's intervention in the materialist moment comes to light. Far from having defended the unity of the self, "Interior man is double, and it is not at all surprising that he should be the object of two sciences that complete each other. Psychology has for its domain sensory consciousness: it knows only what the light of thought casts on sensation; the science of thought in itself, of the light at its source, that is metaphysics."[10] Lachelier contributed to a burgeoning epistemological idiom, one that construed the content of subjective experience according to logic of quality that was rigorous enough to complement psychology's quantitative models.

The materialist moment in ideas of spirit also helps us better understand the inception of psychological research under the Third Republic. The first French psychology laboratory opened in 1889 thanks to Republicans' investment in scientific institutions. A ministerial decree allocated funds to the project as part of the monumental reconstruction of the Sorbonne. The physiologist Henri Beaunis (1830–1921) obtained four rooms in the university, including a demonstration hall and research rooms with time-measuring instruments used to calculate mental duration. Alfred Binet (1857–1911) joined in 1891.[11] He steered the lab's focus toward intelligence measurement, a precursor of modern IQ tests. Yet state investment lagged behind that of other countries.[12] Wundt had established the first modern psychology laboratory in Leipzig in 1879.[13] Across the Atlantic, G. Stanley Hall founded one at the Johns Hopkins University in 1883. Historians have cited spiritualist doctrines' persistence as a reason why experimentation was comparatively late to emerge in France.[14] Psychological research took off—so the argument goes—in industrialized countries where no philosophical establishment cleaved to its institutional authority. But the argument neglects shifts *within* ideas of spirit. Indeed, the two decades leading up to the inception of the Sorbonne's psychology laboratory were an intellectually fertile period.[15] In the following pages, I track the flood of psychological research into France during the 1870s and 1880s, a time when disputes over the nature and scope of mental data preoccupied philosophers, mathematicians, physiologists, and neurologists.

That ideas of spirit came together around a lexicon of *qualitative data* might seem strange from our point of view. The proliferation of computational technologies has left us with an economic and cultural reality in which the concept of "data" goes hand in hand with numbers. Our digital footprints generate ever more data for statistical registries of analysis, organization, and surveillance. In the nineteenth century, however, the nature of data was still up for grabs. Proving that there were facts of subjectivity was crucial to shoring up psychology's basis in physiology. Ribot made the challenge explicit: "We are trying to show that every mental state, whatever it is it is, implies a corresponding physical state."[16] Numbers would make that correspondence legible. Only thus, Ribot argued, would psychology grow into a mature field. "It is still in a qualitative period, and it is uncertain that it will enter so soon a quantitative period, which alone would establish it as science."[17] For spiritualist thinkers, quantitative models offered an experimental point of departure from which to establish equally "mature" qualitative facts. Contemporaneous investigations were conducted in Germany, where the phenomenological structures of first-person experience (i.e., intentionality) were rigorously distinguished from the material objects open to third-person observation.[18] Today, Anglo-American philosophers who explore such questions about the qualitative facets of experience claim to address the "hard problem of consciousness."[19] Spiritualists also countenanced an epistemic gap between the physical and the phenomenal. However, the conceptual stakes confronting their historical context were unique. Thinkers who led the materialist turn in spirit focused their attention on the qualitative evidence of embodied volition.

Quantifying Sensation and Motion

In 1869 a prominent French scientific journal announced, "Thought does not arise instantaneously. . . . It flows with a very noticeable time, one or two tenths of a second, between the moment when the will expresses it by an exterior act."[20] That spiritual activities could be measured with the same precision of physical processes came as groundbreaking news. What had been calculated, in fact, was not thought itself. It was the time that flowed between the moment that a study participant perceived a sensation and the subsequent moment of his motor response. Sensation and motion were the categories employed, first, to decompose thought into measurable units; and second, to make these units analyzable in a laboratory setting. The result was a mechanical picture of the nervous system. But the idea that mental machinery could clock spiritual activities was by no means given. Across the nineteenth century, the development of diverse measurement techniques gave rise to conceptual problems the leaked beyond scientific inquiry.

The physiological foundations of neurological measurement had been put in place thanks to the British anatomist Charles Bell, who had hypothesized that sensory and motor nerves were anatomically and functionally distinct. The French medic François Magendie confirmed that the anterior spinal nerves transmitted motor impulses while the posterior spinal nerves transmitted sensory impulses. Working with a litter of puppies he had procured from a friend, Magendie conducted a series of novel experiments. He first cut the nerve roots along the back of the puppies' spines. They did not respond to pressure. To Magendie's surprise, the animals did move their paws. Incisions to the roots along the anterior side of the spine (which were much more difficult to access with a scalpel) left sensitivity intact, though the puppies' limbs went inert. The two processes, sensation and motion, came to be regarded as basic functions of the nervous system. The German physiologist Johannes Müller subsequently demonstrated that the character of sensation depended on the sensory nerves and not on the physical stimulus.[21] He noticed that the same electric charge produced a buzzing noise in the ears, a pain on the skin, and an acid taste on the tongue. He attributed the diverse sensations to the specific energies of the nerve fibers traveling from the sensory organs to the brain.[22] It had been made clear for nineteenth-century neurophysiology that the self was embedded in a sensory-motor materiality.

Explorations into the electrical nature of the nervous system reinforced the notion that the body was an organic machine. Descartes had proposed that the nerves formed a sinuous network of automatic activity.[23] Machinery ceased to be an exclusively philosophical figure once the Italian physicist Carlo Matteucci (1811–68) observed an electric current flowing through muscular tissue.[24] Further exploration in nervous conductivity made it possible to configure the body into a circuit of inputs and outputs. In *Animal Electricity* (1848), the German physiologist Emil du Bois-Reymond (1818–96) verified that the nerves were conduits of electric charges.[25] He demonstrated that nerves were stimulated only by a change, and not a steady flow, of electric current. The nervous system had a velocity comparable to that of physical bodies in motion.[26]

Just as mechanics had calculated kinetic energy, the nerves' speed was also an object of scientific measurement. This was "physiological time."[27] The Swiss astronomer Adolph Hirsch (1830–91) coined the term, which corresponded to the amount of time that transpired between the reception of a stimulus, its passage to the brain, and the motor contraction in the muscles. The body's machinery introduced a noticeable delay between input and output. In fact, the figure was not lost on Hirsch, who claimed the human body "is exactly like a machine of precision."[28]

Calculating the velocity of nervous transmission became a fixation of the human sciences by the middle of the century. The German physiologist Hermann von Helmholtz (1821–94) discerned the time that elapsed between an electric stimulation and a muscular contraction. He began with frogs. Helmholtz constructed two interconnected electric circuits: the first sent a battery current to a frog's legs; the second measured the current by connecting the battery to a galvanometer. A twitch in the frog's legs would interrupt the current. Helmholtz subtracted the time at which the current was generated from the time it was interrupted. The difference was the speed of the nerve propagation: between twenty-five and forty-three meters per second.[29] The nervous system was shown not to act by instantaneous fiat. Its speed was finite. And it was also variable. Inputs could be modified to calculate different outputs. Building on Helmholtz's studies, Hirsch and his students organized reaction time experiments in which human subjects responded to auditory, visual, and tactile stimuli with a raised hand.[30] The motion indicated that a sensation had been felt. By subtracting the time of the sensory stimulus from that of the motor response, Hirsch compiled quantitative data for a host of sensations. The findings revealed a common denominator: the speed of nervous transmission was twenty-six meters per second.[31] The nervous system turned out not to be a slow machine. Although neither was it breathtakingly rapid.

Emboldened by the advancements of physiological measurement, many psychologists came to believe that spirit could also be measured.[32] For the French physiologist Etienne-Jules Marey (1830–1904), "knowledge of the nervous act, made possible by previous studies of the muscular act, has in turn allowed us to rise up to the study of the psychic act."[33] Reaction time experiments did just that. With his Russian collaborator Nicolas Baxt, Helmholtz measured reaction times to calculate the speed with which visual impressions reach awareness.[34] The Dutch physiologist Franciscus Cornelius Donders (1818–89) and his student Johan Jacob de Jaager went further. They clocked the speed of conscious choice in response to tactile stimuli. The experiments were conducted using a telegraphic device that would emit an electric shock and begin a clock. It stopped when the study participant pressed a key. By subtracting the time of basic reflex reactions (where subjects knew in advance which foot would be shocked) from the time of complex reactions (where they also indicated which foot had been shocked), Donders and de Jaager arrived at the speed of basic mental activity: one fifteenth of a second.[35]

The experiments were suspect. Donders cautioned that unlike the nervous system, spiritual activities "can neither be measured nor evaluated, and we know no unit by which to express sensation, reason, and will in figures."[36] Nervous transmission was divisible into meters and seconds; yet the mental seemed impervious to such measures. The problem was especially trenchant in France, where ideas of matter and spirit pervaded the human sciences. In a widely read report, the mathematician Jean-Charles Rodolphe Radau (1835–1911) declared that reaction time experiments had measured the speed not only of nervous action but also of volition. "An appreciable time flows—one or two tenths of a second—before an idea awakens in the spirit following an impression received by the brain, and the will responds to this idea in bringing about movements of whichever limb."[37] Radau hewed to spiritualist tenets. He claimed that the will dispatched "orders," which relayed "messages" through the nervous system.

These words were ambiguous. What exactly had been measured? Was it an exercise of the will or the action of nerves? Radau's report laid bare the shifting meaning of volition during the nineteenth century. Whereas volition had previously been understood as the spiritual anchor of free will anterior to the body, physiology embedded the concept within the sensory-motor circuitry of the nervous system.[38] Even when it had been adapted to experimentation, volition proved problematic, entangled as it was with individual subjectivity. The French physiologist Albert René found that his measurements of reactions varied according to the intensity of stimuli as well as the kind of research participants.[39] Children, students, and doctors responded at different speeds. For René, reaction experiments confronted an intractable impasse. There were no joints along which to cleanly separate voluntary and neurological activity.

Attempts were made to isolate psychological measurement from neurology. The German psychologist Ludwig Lange demonstrated in 1888 that the speed of conscious awareness was slower than that of motor execution. He claimed the difference turned on the role played by apperception, the psychological act of bringing mental content into focus.[40] Lange measured two simple reactions: in the first, subjects responded while focusing their attention on a stimulus; in the second, subjects responded by directing their attention to the motor response. Lange subtracted the time taken for the muscular reaction, which he assumed to be automatic, from the time taken for the sensory action, which he took to be voluntary. The difference was one tenth of a second: the amount of time that it took to apprehend a sensation. The figure was widely contested. Nonetheless, it fueled the mushrooming ambition to compile a mathematical catalogue of spiritual activities.

Indeed, the accumulation and refinement of mathematical catalogues became a feature of laboratory psychology. Wilhelm Wundt's research set the standard.[41] Equipped with generous funding and five rooms at the University of Leipzig, Wundt opened the Institute for Psychology in 1879, often credited as the first experimental psychology laboratory. His research program in chronometrics came with a dense material configuration. Wires, resisters, switches, telegraph keys, and electromagnets filled the experimentation rooms.[42] Various recording devices included dynanometers, rotating drums, and sphygmographs. Among the technological arsenal of the laboratory, the chronoscope played a significant part.[43] The timekeeping device was powered by an electromagnet attached to weights that rotated needles around a dial. It obtained measurements up to a thousandth of a second. In a dark room called the "reaction chamber," study participants sat in front of an array of contraptions designed to produce visual and auditory stimulations. In one case, a pendulum of slotted black sheets was connected to the chronoscope by a platinum wire. The pendulum would swing to emit a flash of light while simultaneously closing an electric circuit that set the dial's needle in motion. Hidden from the participant, researchers would note the duration between stimulus and response. The experiment would begin with the chronoscope at zero. The pendulum moved the needle, which stopped at .12 seconds, the time it took for the participant to react by pressing a key. The next stimulus would propel the dial until it was pressed again. The dial stopped at .25—so, twelve and thirteen hundredths of a second were displayed on tables divided into columns of reaction types and rows of numerical durations, a visual apparatus that lined the walls of the Leipzig lab.

The ambition to collect more data drove the proliferation of psychological laboratories across Europe and America.[44] Edward Bradford Titchener (1867–1927) dedicated a room to reaction experiments in the laboratory that he founded at Cornell University in 1891.[45] When the Sorbonne laboratory opened in 1889, researchers obtained a chronometer—a more portable, wireless version of the chronoscope—to pursue a research program in reaction times.[46] Although the speed of sensory-motor action had been uncovered, the workings of the brain remained concealed in a black box. The "real time" of cerebral processes could be measured only over a century later with the invention of scanning technologies such as electroencephalography (EEG) and positron emission tomography (PET).

The quest for quantitative data was unyielding in the late nineteenth century. But models built in the image of physics, extended to the nervous system, and elaborated into the domain of spirit, confronted the limits of the mechanical picture of selfhood. Translating the self into a system of inputs and outputs had made it possible to clock psychological activities. Yet, mathematical measurement generated more questions than answers when it came to the connections between spirit and matter. Most pressing was the stubborn problem of volition, which took place in the mysterious interim between stimulus and response. By transposing spiritual powers into mathematical terms, quantitative psychology had configured the will as an opaque machine.

Sensation, Quantification, and Psychophysics

The quantitative revolution took hold in France thanks in large part to Théodule Ribot, the tireless editor and founder of the *Philosophical Review of France and Abroad* (*Revue philosophique de la France et de l'étranger*). Launched in 1876, it was the first French journal dedicated to the bourgeoning field of experimental psychology.[47] Ribot carried out few experiments of his own. Articles were often translations of research collected from laboratories abroad. Ribot used the *Philosophical Review* to foment the swelling passion for precision in psychology.[48]

The journal was meant to be a marketplace of intellectual exchange. Ribot sought to provide philosophers, psychologists, mathematicians, and neurologists with "a neutral terrain where they could produce work, meet, and study each other."[49] His editorial hand, however, was hardly invisible. In the inaugural issue, Ribot announced his commitments in unequivocal terms: "Between natural psychology and metaphysics, there must be a choice."[50] Ribot lambasted spiritualists' claim over psychology, arguing it was a philosophical movement mired in "logical quibbles, imaginary creations, or mystical effusions."[51] In founding the journal, Ribot took immense personal and professional risks. He complained in a letter to a friend, "The *Review* . . . has brought me harassment by the old spiritualists who, at the Institut [de France] perpetually conspire against me."[52] The journal's goal, as he made clear, was to support a physiological psychology that would "study psychological facts from the outside, not from the inside; in the material facts by which they translate, not in the consciousness which gives them birth."[53] In fact, the majority of original articles, 188 in total by 1890, were dedicated to experimental and physiological methods in psychology.[54]

The *Review* came to serve as a key node in a transnational network of psychological research. In 1876, the Scottish psychologist Alexander Bain (1818–1903) launched the British journal *Mind*. In 1881, Wilhelm Wundt followed with the first German psychology journal, *Philosophical Studies* (*Philosophische Studien*). In France, an expansive reading community took shape around the *Review* thanks to its editorial format. As editor, Ribot channeled research, news, and translations from Europe and the Americas. He interspersed articles in philosophy, neurology, and physiological psychology. The journal rapidly became a primary forum where debates and dialogues played out between the spiritualist and scientific study of the self.

Ribot's first article in the journal, as he conveyed in a letter, was "my entry into the German Psychology that I really want to put on display."[55] The piece familiarized French readers with psychophysics and psychometrics, which he believed would ground a new science, "without which, whatever may be said, no research in psychology is possible."[56] Quantification would secure the objectivity of psychological data. Ribot continued to disseminate quantitative models in France with his subsequent book *Contemporary German Psychology* (1879), a project that synthesized the mathematical and material foundations of selfhood.[57]

Psychophysics was an experimental endeavor to apply the logic of quantity to sensory experience. Ernst Heinrich Weber (1795–1898) had developed a preliminary model. The professor of anatomy and physiology in Leipzig conducted experiments aimed at discriminating the threshold for perceiving sensations.[58] Weber had his study participants lift weights. He gradually increased the kilograms until they perceived a difference. For instance, a study participant would lift 10 kilograms. He would then lift 10.5 and 11 kilograms. A change became evident only with the eleventh kilogram. The difference between the initial sensation (of 10 kilograms) and the second sensation (of 11) was what Weber called the "just-noticeable difference." By focusing not on sensations in themselves, but on the interval between them, Weber developed a functional method for measuring discrete sensory thresholds. He inferred that a constant proportion held between changes in the stimulus and the just-noticeable difference. For example, if the same participant lifted 40 kilograms, then he or she would perceive a different sensation when lifting 44 kilograms. The proportion of the just-noticeable difference to the original stimulus ($1/10 = .1$) would also hold for the subsequent stimulus ($4/40 = .1$). Weber applied his original experiments with weights to diverse stimuli, such as the visual perception of the lengths of lines or the auditory sensation of various volumes. His enduring contribution to experimental psychology was the principle of the just-noticeable difference, which linked the psychological domain of sensation to the physiological domain of stimulation in quantitative terms.

Weber measured the difference between sensations, which he represented mathematically as a linear function of stimuli. Was it possible to go further and measure sensation itself? The problem fascinated the German physiologist Gustav Theodor Fechner (1801–87), who had studied under Weber as a young student in Leipzig. In 1828, Fechner retrospectively christened "Weber's Law,"[59] whereby ΔS represents the increase in stimulus added to the initial stimulus S:

$$\frac{\Delta S}{S} = \text{constant}.$$

The law laid the groundwork for psychophysics, which Fechner defined in his 1860 treatise as "the exact science of the functional relations of dependence among body and soul, more generally, between the corporeal and the mental, the physical and the psychological, world."[60] Fechner's ambition was to unite under a single science physics, physiology, and psychology. His motivations, however, were not exclusively scientific. Fechner conceived spirit and matter to be one. A dedicated panpsychist, he believed that both realms were expressions of the same monistic substance. The divergence between spirit and matter hinged on different perspectives. His metaphysical orientation flirted with the spiritualist tradition. "We count as mental, psychological, or belonging to the soul, all that can be grasped by introspective observation or that can be abstracted from it; as bodily, corporeal, physical, or material, all that can be grasped by observation from the outside of abstracted from it."[61] Fechner argued that since the physical was amenable to mathematical measurement, so too was the psychological.

Fechner's ambition far surpassed Weber's. Whereas the latter's work was limited to measuring sensory thresholds, Fechner measured the intensity of sensory experience itself. When a stimulus acted on the nervous system, there would be a proportional effect in the sensation. He was not seeking a causal relationship between stimulation and sensation. His objective was to define the additive structure of sensations.[62] To Fechner's mind, sensation inherently lent itself to measurement since sensory acts could be combined and compared. His mathematical model took as its point of departure Weber's law: $\Delta S/S$. Fechner supposed that within the perceptual threshold of a just-noticeable difference, changes in sensation (ΔE) remained constant. Slight adjustments to the stimulus were perceived to have the same intensity. It followed that sensations within the same threshold constituted stable units that could be quantified. According to Fechner, a functional relationship obtained between the increase in sensation and the increase in stimulus required to produce it.

$$\Delta E = c * \frac{\Delta S}{S}.$$

The formula designated the degree of stimulation corresponding to a unit of sensation. In order to make these units into infinitely small quantities measurable as functions of a stimulus, Fechner derived *E*:

$$dE = c * \frac{dS}{S}.$$

He then integrated both sides of the equation to obtain the relation between the stimulus and the quantity of intensity—the point at which the stimulus is zero and sensation disappears:

$$E = c * \int_{S^0}^{S} \frac{dS}{S}.$$

After deriving the differential of the equation, Fechner arrived at his formula: a logarithmic law that determined the quantity of sensory intensity as a function of the power of the stimulus.[63]

$$E = c * \log S.$$

In other words, the rate at which a sensation intensifies gradually tapers off the more forceful the stimulus. For instance, a constant increase in the wattage of a light produces a progressively smaller increase in the brightness perceived (see figure 1).

Fechner's work prefigured future quantifying practices in psychology. Rather than statistically aggregate the collective responses of a mass of participants, Fechner introduced quantification into the very structure of experimentation. He did not jettison individual responses. They played an integral role. Fechner paired them with numerically graded stimulus series; participants responded to stimuli in binary form (e.g., heavier/lighter; stronger/weaker). These established just-noticeable differences. Fechner's insight was to use them as the foundation of an external model of quantification that would measure participants' inner perceptions.

Could quantities apply *directly* to the structure of sensations? There was a problem: we don't ascertain the quantitative intensity of our own sensory experience. From a first-person standpoint, sensations cannot be compared side by side since they take place at different moments in time. One could not be subtracted from the other. Nonetheless, Fechner claimed that sen-

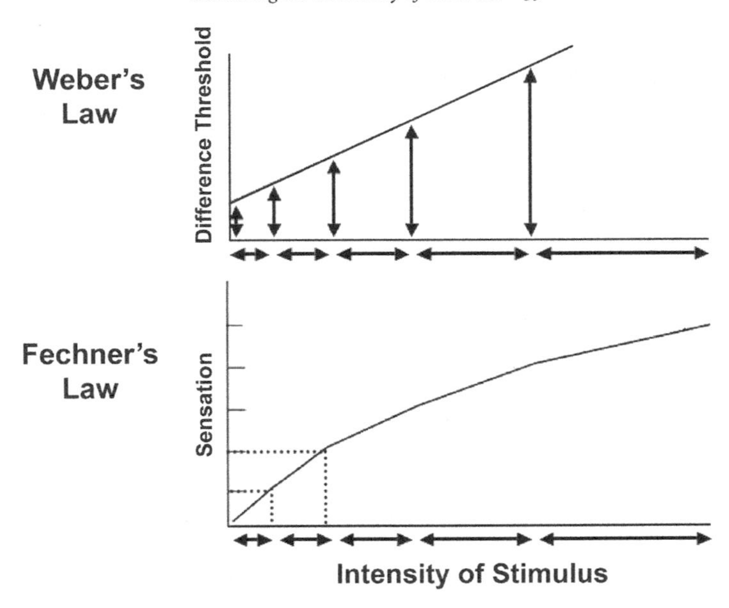

FIGURE 1. The Weber-Fechner law
Source: David Heeger, "Perception Lecture Notes: Psychophysics," http://www.cns.nyu.edu/~david/courses/perception/lecturenotes/psychophysics/psychophysics.html.

sations could be compared in the laboratory as a function of stimuli. To do so required an independent rule. The same holds true for any natural phenomenon. For example, we use the spatial distance between a clock's hands as an independent rule to measure time. Fechner treated the degree of the stimulus similarly. But the stimulus bore only an indirect relation to sensation. A clock would directly indicate the passage of time. The distance between its hands are quantitative indicators that mensurate the uniform quality of seconds.[64] Although stimuli do have equal magnitudes, such as the kilograms of weight or the watts of light, it was not at all evident that sensations also exhibit equality. Fechner took it as his task to demonstrate that sensations have equal degrees of difference, and he did so indirectly—that is, in terms of the magnitude of stimuli. These functioned as the independent rule to compare sensations. By treating just-noticeable differences as basic units of sensation, he conceptualized sensory experience as a collection of perceptual thresholds that accumulate as the stimulus strengthens. Since there was an interval between each increasing sensation at the same time that there was continuity between each increase in stimulation, Fechner could derive the relationship between them.

Fechner's equations spurred a range of disputes. Some claimed to have derived alternate formulae.[65] Others questioned whether psychophysics had actually introduced quantities into the structure of sensations. Rather, it appeared that only the distances between sensations had been quantified. The Belgian psychologists Joseph Plateau (1801–83) and Joseph-Remi-Leopold Delboeuf (1831–96), reconceived psychophysics along these lines. They advanced a bisection method for scaling sensory measurements.[66] A study participant would be presented with two stimuli and asked to identify a third stimulus that fell between them. For example, she or he would select a shade of grey that appeared equidistant between two starker hues. This practical judgment of the distance between two stimuli served as the basis for measuring their respective magnitudes. Plateau and Delboeuf saw their modifications as advancements over Fechner's formula, offering greater precision to psychophysical quantification by means of combination and comparison. Delboeuf in particular stirred interest in the field among the French scientific community, thanks to his popular work,[67] as well as his own quantitative experiments.[68]

Psychophysical research reached a wide audience in France via the *Philosophical Review*. Therein, Delboeuf's contributions appeared regularly.[69] In Ribot's eyes, it had been confirmed that "the fact of consciousness has, like all other phenomena, a precise, variable, and measurable duration."[70] The claim directly contested spiritualists' belief that the immediate data of mental life were not quantities. For Ribot, psychophysics perfected the mechanization of the mind. "The essential has been established: the possibility of measurement."[71]

Quantity and Quality in Sensory States

An anonymously published article in an 1875 edition of the *Scientific Review* (*La revue scientifique*) threw a wrench in psychophysics. The author challenged Fechner's claim to have quantified sensations by extrapolating from Weber's law. "When we don't know at all what signifies the difference between two sensations, how can we speak of the differential of a sensation? What relation is there between a differential and the fact that, by varying the stimulus, we ascertain a moment when the sensation changes? Between the two, there is neither quantity nor continuity."[72] For the author, sensations were not discrete units; they lacked clearly defined boundaries and therefore did not constitute distinct *comparanda*.[73] Moreover, there was no common basis (*tertium comparationis*) on which to compare sensations. He argued that psychophysics had fabricated such a basis by eliding the immediate sensory experience. "Sensation is a phenomenon occurring in us, which we seize in ourselves by its interior side, as it were, and which rebels against any kind of measurement. Without a doubt, a sensation can be more or less intense, but does that suffice to make the sensation a quantity? A quality, beauty for example, can also be lesser or greater. The only *magnitudes* that one can measure directly are those of which one can define equality and addition, and such magnitudes seem to be met only in the domain of abstraction, of pure mathematics."[74] The author drew on the spiritualist distinction between dual dimensions of the self. He claimed that the logic of quantity suited the external dimension. Mathematical models could measure extensive magnitudes underlying quantities. He charged that they were inapplicable to and even adulterated the inner dimension, the felt intensity of sensations. Sensory experience depended on a different kind of magnitude—qualitative in nature.

The author turned out to be Jules Tannery. A young mathematician at the École normale, he was friendly with much of the philosophy faculty there. Although Tannery went on to be one of France's greatest mathematical minds, an intrigue for philosophy also animated his work. His identity as the author became known belatedly—nearly a decade after the article originally appeared—thanks to his older brother, Paul Tannery (1843–1904).[75] Jules valued his anonymity and entrusted Paul with publishing his writings.[76] "If he had not been genuinely horrified by any publicity," a friend wrote, "if he had not rather dedicated himself to not appearing as a creator of metaphysical systems, his name would have been soon known among the public like those of philosophers who are much more fashionable to admire than to read."[77] Despite his distaste for notoriety, Jules Tannery's critique of psychophysics galvanized spiritualists' theoretical engagements with quantitative psychology.

Many in the scientific community took issue with the idea that there was such a thing as qualitative sensory data. Tannery had argued, "We should be skeptical of pompous expressions and a mix of sensations, logarithms, and stimuli." But when it came to an alternative logic for immediate sensory content, "*I confess that I have this logarithm in my heart.*"[78] The notion of quality seemed vague. Ribot claimed that Tannery failed to distinguish questions of metaphysical psychology from those of experimental psychology. "The first are by their nature insoluble and one can say whatever one wants. The second are questions of fact."[79] To Ribot's mind, the quantitative revolution was still inchoate. Technological innovation and methodological refinement would open deeper facets of selfhood up to measurement, including immediate sensory experience. Similarly, Delboeuf argued that the logic of quantity was boundless. "From the moment that one thing is greater or lesser, even beauty, even the pleasure I feel at the sight of a beautiful painting or the hearing of an opera, even if it presents the most obscure of metaphysical reasoning to thought, it can be said that this thing has a size, it has measure, and it is theoretically measurable."[80] A prevailing faith in the virtues of quantification drove many to discount the idea that the subjective content of experience constitutes qualitative data.

Tannery's aim was to articulate the conceptual structure of qualitative data. He took these to be immediate facts, which were epistemologically distinct from the mediate facts of psychological measurement. He reasoned that a different form of validity could allow for the comparison of sensory content. Wundt echoed, "a quality such as beauty can be larger or smaller, but depend on a difference that is not quantitative."[81] Tannery proposed the concept of "heterogeneous magnitudes." Unlike homogeneous magnitudes (e.g., length, area, or time), which allow for comparisons among degrees, heterogeneous magnitudes apply to differences in kind. They apply to singular phenomena that are more than a sum of their parts. Tannery drew examples from aesthetics. To measure the impression of beauty when standing before the *Venus of Milo*, he suggested one could rely on a homogeneous magnitude such as the sculpture's volume, height, or proportion. The sculpture could be compared to other beautiful works such as Michelangelo's *David*. But homogeneity presupposed that the magnitude applied to all the sculptures' parts equally. For Tannery, the *Venus*'s torso, mantle, and face were not equivalent components; they inhered in the sculpture's heterogeneous composition. "Certain works impress us more than others. But between these diverse impressions, is there not a difference of more and less, but *a difference of nature*?"[82] His claim was that that a heterogeneous magnitude captured the affective impression of the sculpture in its entirety. A world stood between the *Venus* and *David*. As a consequence, comparing the sculptures' beauty would depend not on a quantitative calculation, but on a thick description of the their distinct qualities.

The aesthetic examples were especially fitting since Fechner had claimed to calculate psychophysical standards of beauty in his *School of Aesthetics* (*Vorschule der Aesthetik*) of 1876. Against philosophical aesthetics, which drew concepts such as pleasure, taste, and satisfaction "from above," Fechner sought to construct an experimental aesthetics "from below." He calculated aesthetic judgments in three studies. In the first, participants selected artworks that they found most pleasurable among a range of options; in the second, participants made art objects that they found to be either the most or the least pleasurable; for the third, Fechner compiled statistics of the forms manifest most frequently in artworks. Using catalogues of European art museums as his guide, he proposed a "golden rule" of proportionality.[83] Fechner identified the most pleasurable rectangular proportions evident in sculptures from antiquity. If a torso were too long or legs too short, the figure would violate the golden rule—a rule that Fechner believed to represent the "divine" mathematic standard of beauty.

Tannery developed his conceptual lexicon, and the notion of heterogeneous magnitude, alongside Émile Boutroux. The two were students at the École normale from 1865 to 1868. In Boutroux's eyes, Tannery "was interested, with his open spirit, in my efforts to understand his scientific reasoning. Penetrating his soul, mixing my ideas with his was one of the most profound joys he had given me to taste."[84] Both began their academic careers teaching in a lycée in the northern French town of Caen in 1871. In fact, Boutroux was a fixture in the Tannery family.[85]

The Tannery-Boutroux correspondence reveals a personal and probing investment in the methods of scientific inquiry. During his doctoral studies, Boutroux drafted a map of the sciences, a preparatory schema for his thesis, *The Contingency of the Laws of Nature*. He and Tannery discussed how to organize the sciences in two categories: quantity and quality. The concepts lent structure to the spectrum of scientific inquiry. At one end of the spectrum, mathematics examined pure quantities. At the other, philosophy addressed pure qualities. Boutroux spent 1873 arranging the intermediary sciences, from physics and chemistry to biology and psychology. "I would like to determine the species of quantity of which each science studies the permanence. There would be the formula of kinetic energy (mv^2), weight, etc."[86] Despite his esteem for scientists, Boutroux was hardly one himself. He bashfully asked Tannery, "You could without a doubt give me the formulae responding to my question."[87] According to Boutroux, mathematics, physics, and chemistry relied on a quantitative logic to determine the relations among natural phenomena; but they were also imbued with indeterminacy. "Generally what remains is the possibility of there being *different* uses for a given quantity while holding onto the permanence of the amount. For example, the antecedent 4 works equally well as a consequent of 2 + 2 or 1 + 3. I would like to find something analogous to the example in all of the sciences. This chain of several solutions, equally possible and legitimate from the scientific point of view, would be an element of chance contained in nature, studied with just the illumination of the understanding."[88]

The "chance contained in nature" was reflected in the qualitative facets of science. For Boutroux, quality was most pronounced in what he called the "moral sciences"—psychology, sociology, and philosophy, each of which took free human beings as its object of inquiry. But quality also manifested throughout the spectrum. "In assigning it an appropriate place in the object of each science we could explain what in each of these objects would remain unexplained."[89] For Boutroux, the logic of quality gave meaning to the contingent nature of spiritual creativity. In dialogue with Tannery, Boutroux claimed that scientific laws were human creations, which necessarily integrated the constitutive instability of their creators.

What motivated Boutroux to conceive of contingency as the hallmark of quality? Why were qualitative concepts so indispensable to understanding human creativity? Tannery drew Boutroux's attention to psychophysics and psychometrics. Both relied on an additive model of the mind that divided sensory experiences into discrete and comparable units. "We know that hunger can be greater or lesser, and from there we represent it as an intensive quantum that could theoretically be measured absolutely like heat or light. But that is nothing. In effect, one can't feel a hunger in one given moment, and one can't compare with oneself the sensations of two different hungers."[90] According to Tannery, sensory states could not be experienced side by side since they inhered within the self, a self whose interiority was structured by the will's dynamic continuity. As Boutroux claimed in his subsequent lectures, "Qualities cannot be set outside of one another like material things. It is impossible for us to say where one ends and another begins: they are insuperably complex and fluid."[91] For Tannery and Boutroux alike, qualitative states issued from the inner life of spirit—they constituted the immediate content of experience—whose free activity unsettled mechanistic models of selfhood.

These thinkers intervened in a scientific atmosphere where the mechanic figure of mental inputs and outputs circulated widely. Exemplary was the work of the Leipzig laboratory, where the psychologist Martin Trautscholdt claimed to measure the duration of mental association.[92] Trautscholdt would stand in front of a study participant, utter a monosyllabic word, and begin timing with a chronometer. Once the participant recognized the word, he would hit a button to stop the time. The data were aggregated and subtracted from simple reaction times to nondescript sounds. Trautscholdt concluded that the difference was .0727 seconds: the time it took for a memory to enter awareness. The laboratory produced further results: the time taken to recognize the quality of sensations (e.g., that a tone is deep or that a color is red); to memorize sounds and letters; and to perform simple manual tasks. Underlying the lab's work was a conviction that mental activities could be broken apart and reassembled; that psychological qualities could be measured on the basis of physiological quantities. Further measurements were conducted on the duration of mental associations under conditions of fatigue, exertion, and intoxication.

There was little consensus that volitional activity could also be measured. The physician Charles Féré (1852–1907) claimed to calculate the magnitude of effort using a dynamometer.[93] The instrument was initially devised to gauge the torque of engines. It also came to be used in psychophysical calculations. Gripping the dynamometer registered the hand's pressure, which served as the basis for Féré's calculations of muscular force. According to the philosopher Alexis Bertrand (1850–1923), the procedure was dubious. "Passing from the effect to effort, from quantity to quality, from the extensive to the intensive, that would indeed be the transcendental paralogism of physiology."[94] Similarly, Adolphe-Moïse Bloch (1842–1920), a doctor at the Asile de Vincennes hospital, believed that volition was too unstable to calculate. He agreed that neurological activities were open to measurement, but "they cannot serve to measure the time of sensitive transmission because they bring into play a factor: *the transformation of a sensation and volition*, independent of the length of sensory nerves."[95] A strict parallel did not obtain between nervous and volitional activities. Bloch's work compared the speed at which sensory impressions traveled to the brain. Vision was the fastest; audition was .0138 seconds slower; and touch was another .0476 seconds slower than vision.[96] These quantitative models, he claimed, could not be applied to mental acts of perception. "Voluntary responses contain a psychic element whose duration, and manner of being, is unknown: the transformation of perception into volition. This element varies from one sensation to the next, without us knowing why."[97]

For Boutroux and Tannery, the "psychic element" injected contingency into experimental practices. They argued that volition qualitatively transformed between stimulus and response. According to Tannery, "nothing can prove that all the anterior phenomena, in two different cases, would neither be the same as the sensation felt by the subject, nor that the consequences that follow would be the same."[98] Resultantly, volitional activity eluded the precise calculations of reaction time experiments. For Boutroux, exercises of the will could not be likened to mechanistic outputs. "A psychological consequent never finds in the antecedent its complete cause and all-sufficing reason."[99] He reasoned that the will executed a higher order of determination, a self-determination according to which the conditioned surpassed its conditions. Herein lied the metaphysical connection between contingency and quality, as Boutroux argued in *The Contingency of the Laws of Nature*:

> To consider quantity with relation to a homogeneous quality, or to leave quality altogether out of account, is to place oneself outside the conditions of reality itself. Everything that is possesses qualities, and consequently participates in that radical indetermination and variability which belong to the essence of quality. Thus, the principle of the absolute permanence of quantity does not apply exactly to real things: these latter have a substratum of life and change, which never becomes exhausted. The sin-

gular certainty presented by mathematics as an abstract science does not authorise us to look upon mathematical abstractions themselves, in their rigid monotonous form, as the exact image of reality.[100]

Boutroux's claim that mathematical models were abstractions unmoored from "life and change" was not meant to condemn quantification *tout court*. He sought to exceed its limitations and advance a concept of qualitative data that he and Tannery had originally formulated in the lexicon of heterogeneous magnitudes. In the final chapter of *The Contingency of the Laws of Nature*, Boutroux tackled the logic of quality governing spiritual interiority. The will, the energetic source of contingency, cut across the multitudinous planes of selfhood. "Even in the moral order of life, beneath changing externals, there are strata ever more and more solid. Beneath the disposition of the moment is individual character; beneath individual character are the manners and customs of the time; then follows national character, and, finally, human nature itself."[101] He reasoned that psychologists' mechanistic picture had neglected the flux of mental dynamism, the permanent potential for humans to radically transform their habits, cultural mores, and social institutions. Boutroux went on to suggest that psychology remained incomplete so long as it studied mental facts from the one-sided viewpoint of quantity. Absent the tools of qualitative description, "scientific concepts, intelligible as a measure of reality, will lose all meaning if we insist on the measure finally measuring nothing but itself."[102] Only spiritualism would bring psychology to completion. Science "will leave spirit subsisting, and with it the possibility of a spiritualistic metaphysics."[103]

From Spiritual Interiority to the Social Self

Endeavors to articulate a qualitative logic of immediate data did not only address inner states of experience. The materialist turn among French ideas of spirit also found in quantitative models the tools to illuminate the social nature of selfhood. Measurements of mental duration prompted spiritualist thinkers to reexamine the parameters and setup of the laboratory setting. Experimental practices were not neutral. Psychometric and psychophysical measurements relied on a particular configuration of the research subject, a configuration that elicited an externalized form of temporality. For spiritualists, quantitative psychology drew the self, as it were, outside itself. What had been calculated was a social span of time: a duration shared with others, but qualitatively distinct from the subjective data of temporal experience.

Bergson argued along these lines in his first book, *Time and Free Will: An Essay on the Immediate Data of Consciousness*. He argued that the subjective experience of time—what he called "lived duration"—eluded the grasp of mathematical measurement. Personal and social experience, although they originated within the self, did not share a uniform temporal structure. Unlike the discrete mental acts clocked by quantitative psychology, lived duration exhibited an indivisible continuity composed of seamless transitions. In the eyes of readers, Bergson's debt to his fellow spiritualist thinkers was legible. According to Jacques Chevalier, "In the *Essay* . . . what appeared completely new for the era was not the critique of Fechner's theory of intensity, since we had all read Tannery; it was his conception of duration: for we had all relied on Kant, and on the a priori forms of the sensibility."[104] In the first *Critique*, Kant had characterized time as the linear form of inner states: "We reason from the properties of this line to all the properties of time, with this one exception, that while the parts of the line are simultaneous the parts of time are always successive."[105] By decomposing time into separate instances that were juxtaposed successively, Kant had established an epistemological framework for the enumeration of psychic states in discrete quantities. To Bergson's mind, Kant had misunderstood the inner experience of time by presenting it in spatial terms. "Kant's great mistake was to take time as a homogeneous medium. He did not notice that real duration is made up of moments inside one another, and that when it seems to assume the form of a homogeneous whole, it is because it gets expressed in space."[106] For Bergson, we think spatially because we live socially. We thus become accustomed to thinking of time in the image of outer states. But we're also capable of resisting our social inclination and reflecting inward on the fluid passage of lived duration that interweaves in a heterogeneous flux.

I am suggesting that Tannery's notion of heterogeneous magnitudes and Boutroux's work on qualitative contingency offered a conceptual idiom in which Bergson articulated his account of lived duration.[107] In fact, Boutroux was Bergson's professor at the École normale. He reflected, "Boutroux had the reputation of not admitting that philosophy could have a point of departure other than Kantianism."[108] After taking the course a second year, the young Bergson warmed up to his professor. "For this grand adversary of 'scientism' actually respected science more than anyone else."[109] Boutroux often encouraged his classes to enroll in scientific courses as well. Among his archives was a letter encouraging a student to write a thesis on "the experimental method and its application to psychology."[110]

Bergson went beyond fellow spiritualists in explaining the social conditions of quantification. In *Time and Free Will,* Bergson presented space and time as dual tendencies. The former emerges from our coexistence with others; the latter is immersed in our mental interiority. He claimed that each accords with a distinct logic: space is quantitative; time is qualitative. To illustrate the distinction, consider two representations of a willed activity such as running.[111] With quantitative concepts, we could represent the runner traversing a determinate length of space. Each step of his or her path would occupy a point along a line. The points would be measured in uniform units that allow us to compute the runner's average speed. Say the runner completes five kilometers in thirty minutes. The average speed would be ten kilometers per hour. For Bergson, this representation is spatial (or better, extensive). It breaks apart the runner's movement into homogeneous quantities that facilitate mathematical measurement. With qualitative concepts, we could represent the runner temporally (i.e., from within). It would look quite different. The entire process—from his or her bursts of energy through fatigue and, ultimately, exhaustion—would comprise an uninterrupted stream of intensive states. Apprehended from the point of view of the runner's immediate experience, his or her movement constitutes a lived duration continuously interwoven with acts of exertion.

Space and time were indispensable categories for spiritualists' understanding of the dual structure of selfhood. The philosopher Victor Egger (1848–1909) drew a sharp division between the extended and the unextended. Whereas extended phenomena could be located in the body (thus laying claim to physiological facts), unextended phenomena were uniquely temporal. Egger reasoned that the latter exhibited a durational structure: "I am a pure succession; facts which are neither extended nor localized, but have their own duration of antecedents and consequents—these are my facts."[112] There was a precise conceptual division of labor. The spatial and the temporal were distinct orders of experience corresponding to the respective logics of the material and the spiritual. "To explain psychological facts by physiological facts supposes a middle term half-extended, half-unextended. . . . And there you have it: no discovery could establish a link between the brain or its function (both being extended) and unextended thought."[113] In a widely read debate in the *Philosophical Review,* Charles Richet (1850–1935), the eventual 1913 winner of the Nobel Prize in Physiology, criticized what he took to be Egger's "spiritual game," full of "vague reasoning and metaphysical conjectures."[114] Richet placed his complete trust in the progress of experimentation. "Leaving the human sciences to logicians who disdain the experimental method . . . is contrary to progress."[115] For Egger, such methods were bound to spatial categories, which failed to grasp the fluidity of mental interiority.[116] On the back of his personal copy of the debate in the pages of the *Philosophical Review*, Egger jotted some notes to himself. He titled them, "questions that a serious adversary would pose to carry on the discussion."

1. Is pure independent psychology, according to the introspective method, a science; that is to say, does it surpass the description of facts and explain them? Does it have laws?
2. The psychology of phenomena is hardly constituted. But even if it were, would the psychology of sensations still not be enough for physiology?[117]

The notes offer an intimate glimpse into the philosopher's aspiration to make spiritualist concepts appear rigorous in the eyes of his scientific peers. Egger sought to set the temporal structure of mental phenomena on a firm epistemological foundation. Duration could thus be a source of positive facts consonant with the spatial facts of physiological psychology.

Pursuing a parallel line of thought, Bergson posited facts of duration. He did so by contrasting their qualitative composition with the spatial form of psychophysical measurement. "The mistake which Fechner made . . . was that he believed in an interval between two successive sensations S and S′, when there is simply a *passing* from one to the other and not a *difference* in the arithmetical sense of the word."[118] According to Bergson, psychophysics presupposed that sensations could be juxtaposed as if they occupied a uniform medium. The presupposition was not peculiar to psychophysics. In Ribot's words, psychometrics "made us comprehend why consciousness consists of a discontinuous series of states separated by short intervals."[119] Bergson responded that the social conditions of experimentation had brought sensations under a spatial mode of representation. As a result, sensations assumed the form of immobile units. By contrast, "Pure duration is the form which the succession of our conscious states assume when our self [*moi*] lets itself *live*, when it refrains from separating its present state from its former states."[120] The most significant sensation was that of freedom. For Bergson, acting freely is not an all or nothing wager. Volition admits degrees. Whether we exert our will strenuously or effortlessly, the infinitesimal gradations of lived duration populate the felt impressions of experience—that is, spirit.

Did Bergson believe that one couldn't draw *any* comparisons between moments of lived duration?[121] Must we adulterate the immediacy of temporality by distinguishing past and future, lesser and greater, near and far? Bergson did not contend that lived duration admits no differences. At stake was what counted as a difference. Bergson framed the problem such that space and time hinged on distinct varieties of multiplicity. On the one hand, spatial representations constitute what he called "discrete multiplicity." On the other hand, there is a form of multiplicity specific to temporal experience, a "form [that] duration assumes when the space in which it unfolds is eliminated."[122] Bergson called this "qualitative multiplicity." It is a mode of representing intensive states. These do not differ externally from each other, but instead differentiate internally from themselves.[123] For Bergson, discrete multiplicity and qualitative multiplicity lent structure, respectively, to the logics of numerical data and immediate data.

Consider again the runner. He or she might occasionally reach a comfortable stride or at other times struggle to persevere. The difference between a casual pace and a furious exertion is intensive; it depends on a modulation of effort. Different levels of effort are not discrete multiplicities. They express fluctuations of the will. For Bergson, differences in intensity are qualitative multiplicities: they are inseparable from the self, inhering in the same duration, a duration in which volition modulates itself (to wit, self-differentiation).[124] The immediate data of duration include a sensory element neglected by quantitative psychology: the sensation of transition from one state to another. For instance, slowing down is felt with relief after speeding up. "We should therefore distinguish two forms of multiplicity, two very different ways of regarding duration, two aspects of conscious life. Below homogeneous duration, which is the extensive symbol of true duration, a close psychological analysis distinguishes a duration whose heterogeneous moments permeate one another; below the numerical multiplicity of conscious states, a qualitative multiplicity; below the self with well-defined states, a self in which *succeeding each other* means *melting into one another* and forming an organic whole."[125] According to Bergson, mental states retain the trace of past states.

In terms strikingly similar to Bergson's, Alfred Fouillée argued that quantification had distorted the immediacy of temporal experience. He mounted a critique of mathematical models that "divide the indivisible, render homogeneous the heterogeneous, immobilize the mobile, and render static the dynamic."[126] Fouillée articulated a comparable logic of quality that would capture both the conceptual structure and affective density of volitional activity. The similarity of his project to Bergson's was not lost on Fouillée's grandson and biographer Augustin Guyau: "in opposing the dynamic to the static, time to space, both want to safeguard liberty and novelty in the world."[127] In a comically hostile manner, Guyau hastened to note the lone difference between the spiritualist thinkers: "Fouillée stays in the domain of experience; Mr. Bergson, believing to stay there, returns to the conception of an elusive and incomprehensible liberty, analogous, despite his intentions, to Kant's noumenon or Schopenhauer's thing in itself."[128] Although aware that Bergson's fame overshadowed his own, Fouillée maintained an amicable relationship. "What a philosopher, Bergson, entirely into science and philosophy," Fouillée reflected, "working with such an upstanding awareness and never set back by fatigue!"[129] In fact, the two just missed each other at the École normale supérieure, where Bergson entered as a student in 1878, three years after Fouillée's departure.

Unlike Bergson, Fouillée came to philosophy as an outsider. Saddled by financial hardship after his father's slate quarry business went under, Fouillée did not have enough money to attend the École normale. His route to the French professoriate was unconventional. He worked for the Dictionnaire des contemporains, a collection of contemporary thinkers' work run by the lexicographer Louis Gustave Vapereau (1819–1906). One of Vapereau's editors noticed the young Fouillée's sharp mind. With his encouragement, Fouillée sat the agrégation in philosophy in 1864 and took first place. His success drew the attention of the Parisian faculty. Cousin counseled Fouillée to enter essay competitions organized by the Académie des sciences morales et politiques. His submission on the theory of ideas in Plato won him the prestigious Bordin prize. In the conclusion, he argued, "all oppositions among diverse philosophical schools, diminishing bit by bit with the progress of analysis, tend to disappear into a comprehensive and synthetic doctrine."[130] This planted the seed of Fouillée's lifelong philosophical project to bring about a "conciliation" of science and metaphysics.[131] In his 1872 thesis, *Liberty and Determinism* (*La liberté et le déterminisme*), Fouillée argued that the two ideas do not conflict. He set the spiritualist commitment to free will within the determinate order of nature.

Fouillée contrasted his conciliatory project, aimed as it was toward the future, with Victor Cousin's eclecticism, which Fouillée took to be fastened to the past. His aim was not just to dissolve philosophical oppositions. Fouillée saw himself as a philosopher who stood on solid scientific footing. He sought to build a higher idea of spirit premised on embodied volition: "Where science ends, metaphysics begins, and above all, the metaphysics of action."[132] In fact, Ribot admired Fouillée as "the most brilliant mind," who "like the rest of his school, fully accepts the data of science on condition of subordinating them to the moral point of view, which contains the true ground of things."[133]

Fouillée drew on concepts of spirit to expand psychologists' understanding of selfhood beyond the limits of quantification. Like Bergson, he reasoned that numerical models relied on a spatial representation inherited from the Kantian conception of temporality. That mode of representation "always gathers together time and space on the same line, as if what applies to one would also apply to the other."[134] According to Fouillée, quantitative psychology occluded a more expansive realm of subjective interiority. "Life, sensibility, even consciousness is not a single and indivisible thing as imagined by the traditional spiritualism: it is susceptible not only to multiple directions, but diffusion, concentration, transmission, and displacement."[135] The logic of quantity brushed only the surface of selfhood. Such was especially the case in the laboratory setting, where mental activities were measured as if they were limited to processes within the brain.[136] Characteristically, the physiologist Étienne-Jules Marey (1830–1904) claimed, "the *cerebral act has a duration*. It has ultimately been assured that this duration increases in proportion as the psychic act becomes more complicated, and that it decreases when the intellectual operation is simplified."[137] To Fouillée's mind, what went missing in laboratory experimentation was an appreciation of qualitative data, which occupy the interstice between stimulus and response.

Despite his criticism, Fouillée believed that the sensory-motor orientation of physiological psychology offered a springboard to imagine a more dynamic spiritualism in contact with the material world. The crux of his claim was that the mind is not a self-enclosed container of ideas that represent the world in conceptual form. Ideas are powerful forms of action. He called these *idées-forces* (which could be translated—albeit awkwardly—as "idea-forces"). In a programmatic essay, Fouillée argued that idea-forces represent a conciliation of physiological psychology and the spiritualist anthropology of selfhood. The "*motor* power of the *idea*," he claimed, was the "middle term between the two."[138] These follow the centripetal impulsion inward (sensory reception) and the centrifugal impulsion outward (motor response). Fouillée said that the notion of idea-forces, central as it was to his lifelong work, had developed from his thesis on liberty and determinism, in which he argued that "liberty constitutes itself in being thought, that it is as such a real force at the same time that it is ideal, that it is its own means of efficacy."[139] For Fouillée, idea-forces are expressions of volition and thus enter into resistance with a shared social and biological world.

Furthermore, idea-forces relied on the insights of quantitative psychology, which had verified the imbrication of embodied motility and spiritual interiority. In particular, reaction time experiments "prove that the operations of consciousness are all tied up with movements in a resistant milieu. If there were not at the same time a competition and conflict of forces when we try to think, compare, and desire, we would see no reason for these acts not to burst forth instantaneously, like the omnipotent *fiat* of a solitary self sufficient to himself."[140] In other words, volition takes time because it involves far more than a decision. For Fouillée, the measurements of mental duration reflected the amount of time that idea-forces spent reaching out, affecting, and manipulating the material world. "If there is a resistant milieu, there is movement in space and not only a change in time."[141] Missing in the laboratory setting was a rich description of idea-forces' underlying engine—what Fouillée called *appétition*. Appétition "pushes a living being toward action in view of the satisfaction of a need or a desire or the realization of an end."[142] Fouillée's understanding of appétition could be likened to Spinoza's *conatus*: a being's innate inclination toward self-enhancement. Fouillée explicitly drew on Darwin's account of the competition for life.[143] Organisms express their vitality in the ongoing struggle to minimize pain and seek out well-being. "The appetitive action that constitutes life and the will is constant, without actual intermittences."[144] To comprehend the immediate data of appetitive action required an examination of volition in its fluidity. Otherwise, as quantitative models measured the disjointed instants of time passing between inputs and outputs, "Consciousness becomes a spectator, a witness, an eye contemplating over things all the while it can do nothing; less than that, a simple reflex that throws a momentary flash on a genuinely obscure and unconscious background."[145] In its analysis of individual acts, mathematical psychology lost sight of the broader social and material context of spiritual activity.

Fouillée and Bergson shared a diagnostic approach to quantitative psychology. They did not argue that mathematical models were wrong. Much less were they mere abstractions that scientists had imposed on the mind. Instead, these spiritualist thinkers argued that mental chronometry, psychophysics, and psychometrics lent mathematical form to the sociality already inhering in the self. Our social existence makes possible the calculation of matter and motion in a shared space. "We find it extraordinarily difficult to think of duration in its original purity," Bergson wrote. "This is due, no doubt, to the fact that we do not *endure* alone; external objects, it seems, *endure* as we do, and time, regarded from this point of view, has every appearance of a homogeneous medium."[146] Although quantitative psychology relied on spatial categories, Bergson reasoned that it was blind to the genesis of those categories within the self. For Fouillée, inherent in idea-forces was the power of self-differentiation, a genetic power to produce the dual dimensions of exteriority/interiority, space/time, quantity/quality: "they are not two disparate realities that could be indifferent to one another, nor two 'aspects' of which one, the mental, would be the epiphenomenon of a phenomenon; rather it is *the same reality* in the process of development that divides itself by the diversity of means to seize it."[147] For Bergson and Fouillée alike, there was not identity at the core of the self but difference.

Bergson added a metaphysical twist. He acknowledged that it is perfectly natural to succumb to our social tendencies and deploy compartmentalized, interpersonally manageable quantitative data. But he also thought that qualitative data are — in a strict sense — unnatural. We come to comprehend them by resisting our social disposition to quantify. To remain within oneself and apprehend the immediate facts of lived duration requires a heightened concentration, what Bergson characterized as a "strenuous effort of reflection."[148] Mental exertion thereby folds back on itself. The gap between subject and object breaks down, leaving a field of experiential content. In *The Creative Mind* (*La pensée et le mouvant*) (1934), Bergson called this activity intuition: "To think intuitively is to think in duration." He clarified, "The habitual labor of thought is easy and can be prolonged at will. Intuition is arduous and cannot last."[149] This theme of Bergson's thought is the subject of chapter 5. For now, I want to highlight that the effort involved in self-reflection involves a thrust against one's social predilections. By turning against sociality, the self enters into and attains absolute knowledge of its inner creativity. This was the paradoxical metaphysics that underwrote the materialist turn in spiritualism. Bergson showed that qualitative data lay at the foundation of selfhood; yet to return, comprehend them, describe their contours, and articulate the structure of immediate experience, spiritualism took as its point of departure the numerical data of quantitative experimentation.

It was only in the final decades of the nineteenth century that psychological research achieved institutional recognition in France. Higher education integrated quantitative methods when Ribot took the first chair in psychology at the Sorbonne in 1888. The next year, the first International Congress of Psychology took place in Paris on the occasion of the Universal Exposition of 1889.[150] It had already been set in motion by the Society of Physiological Psychology, founded in 1885 thanks to the organizing work of Ribot. His requirements for membership, as he conveyed in an invitation, were straightforward: "no metaphysics and meetings once a month."[151] Yet metaphysics, as I have argued in this chapter, was an inexorable outgrowth of the very mathematical models designed to contain it.

Once a laboratory for experimental psychology opened at the Sorbonne in 1889, psychological quantification blossomed in new domains. Alfred Binet devised scales for intellectual measurement, showing the variation of intelligence with age and developing some of the earliest personality tests.[152] At the Villejuif Hospital, the psychiatrist Édouard Toulouse (1856–1947) organized a second psychology laboratory.[153] Beginning in

1901, his research team studied the relations of genius and madness, a project that established mathematical rigor for calculations of individual differences. The emergence of public sanitation programs, the expansion of factory production, and bureaucratic streamlining of the military gave rise to new demands for mass metrics of social filtration. Psychodiagnostics constructed measurements for abstract, mechanical, and social intelligence. And psychotechnics generated tests to measure workers' aptitudes.[154] By the early twentieth century, the authority of numbers pervaded France.

Far from having reproached the quantitative revolution, the materialist moment in concepts of spirit grew out a creative appropriation and reinterpretation of psychological literature. Psychometrics and psychophysics spurred spiritualist figures to shore up the factual bases of selfhood. Tannery, Boutroux, Bergson, Egger, and Fouillée contributed to overlapping accounts of qualitative data, accounts that bestowed an empirical authority on subjective interiority. Numbers were not the only form of data. Scientific validity was not exclusive to the measurement of matter. A validity unique to immediate experience was equally authoritative.

The figures in this chapter advanced seemingly idiosyncratic lexicons. Heterogeneity, contingency, duration, and appétition: these concepts emerged out of a sustained interrogation of psychologists' quantitative results. The materialist current in French spiritualism converged around a shared commitment to make the dynamic fluidity of the will, and not only its numerical trace, into a rigorous object of inquiry. Moreover, exploration into the spiritual conditions that made quantification possible laid bare the sociality of the self. This evolving preoccupation with the social dimensions of selfhood propelled the turn to matter in French spiritualism. It came to be increasingly evident that spirit was not confined to the realm of interiority, but entered into and impinged on our shared material world.

✷ CHAPTER 3 ✷

Science and Spirit in the Classroom

Before Parisian crowds filled his celebrated lectures at the Collège de France, Henri Bergson was a little known yet tireless philosophy instructor in a provincial lycée (the French equivalent of high school). In September 1883 he came to the town of Clermont-Ferrand, where he taught at Lycée Blaise Pascal. Bergson opened the first lesson on psychology, the centerpiece of the course, with an introduction to neurophysiology. He proceeded to critique the reductionist implications of the emergent science: "If sadness were in the heart, if thought were in the head, it would occupy a place there, which by dissecting one could end up finding on the end of a scalpel. . . . But thought does not reside in the brain."[1] The lesson reflected the central argument of his doctoral thesis, *Time and Free Will: An Essay on the Immediate Data of Consciousness* (1889). "Thought is nowhere, it doesn't have a place without space, it a has a duration like a feeling, but it is not extended."[2] Writing and teaching often coincided, as notes written by students in the course make clear. But the lesson was not entirely Bergson's own. It revealed an ideological cleavage at the heart of philosophy's public mission under the Third Republic. Bergson was an independent thinker indebted to the nation's intellectual heritage as well as a civil servant tasked with inculcating secular and scientific values to future generations of French citizens. The psychology section of the lycée curriculum offered philosophy instructors a forum in which to broach conversations between the young brain sciences and spiritualism.

Between 1874 and 1902 the French Ministry of Public Instruction introduced lessons in experimental methods, psychopathology, and neurology into the lycée philosophy course. I argue in this chapter that these educational materials were critical to the institutional contexts that sustained the materialist moment in French spiritualism. During this period, the spiritualist orientation of the curriculum served a particular goal. The philosophy course was mobilized to promulgate nascent brain science; and spiritualist ideas came to serve as a pedagogical vehicle that would steer its cultural dissemination.

Bergson's time as a lycée professor is especially illuminating. The philosophy course, and its psychology section in particular, provided Bergson with the opportunity to integrate developments from neurophysiology into his early writings. My claim is not innocent.[3] Bergson acknowledged, "I took as a maxim at the Collège [de France] not to bring my current research into the direct subjects of my courses. That was even more the case for boys sixteen years old."[4] It is certainly true that Bergson's courses were not expositions of his ideas. I want to suggest, nonetheless, that secondary education, and specifically the experience teaching psychology in the philosophy course, throws into stark relief one of the institutional settings in which Bergson emerged as the most successful representative of spiritualist materialism.

Philosophy was originally introduced in secondary education under Napoléon in 1809. That same year the Imperial University established the École normale supérieure at rue d'Ulm in Paris to train young professors. Access to the terminal course in philosophy remained a privilege open to the sons of the bourgeois elite for most of the nineteenth century. It has remained the terminal course of lycées to this day. *Philo,* as the final year is called, was originally designed to synthesize a scientific and humanistic education. The curriculum followed the standardized program released by the Ministry of Public Instruction. After opening with a brief introduction to philosophy, the first section of the program was psychology, followed by logic, metaphysics, ethics, and the history of philosophy. Psychology was the centerpiece of the program since Victor Cousin had introduced the section in 1832 while head of the ministry. Students were taught psychology in order to understand the self. For Cousin, psychology was an introspective study inimical to empirical verification. It paved the high road to ontology by examining the faculties of consciousness—sensation, reason, and the will. This brand of eclectic spiritualism had "a good claim to be the national philosophy of France."[5] It was enshrined in the lycée curriculum for much of the nineteenth century. Cousin's monopoly over the education system enforced his eclectic approach to psychology by fashioning the section's lessons around the self, whose identity and freedom independent of sensations served to anchor introspective inquiry.[6] To this day, psychology endures under the title "the Subject," which all students study before sitting the *baccalauréat,* the official exit examination.[7]

Following Cousin's death in 1867, the 1874 program brought about a significant rupture with the old spiritualism.[8] The psychology section incorporated physiological and experimental content. In the early years of the

Third Republic, the philosophy course became a social laboratory. It set advancements in the sciences within a pedagogical framework sensitive to the moral and nationalist imperatives of a fragile democracy.

The scientific turn in philosophy instruction was part and parcel of the government's modernizing initiatives, which gained traction after a majority of Republicans were elected to the French parliament in 1880. The state took on expansive powers to remake French society. It reached down into people's lives through wide-ranging agents: postmen, bureaucrats, policemen, railway employees, and—most of all—teachers. Schools were the primary engine by which the state made the nation (and not the other way around). Teachers promoted hygiene, expanded literacy, and homogenized French language instruction—elements integral to France's modernization.[9] Republican reformers invested in primary schooling to mold a modern citizenry and emphasized secondary schooling to cultivate an intellectual elite.

More specifically, the scientific turn in philosophy served to promulgate Republicans' *laïque* reforms. *Laïcité* is a peculiar variety of French secularism. Defined negatively as the curtailment of religious influences over public life, laïcité found its positive content in scientific instruction. Education "passed on first and foremost a cult of science," according to Philip Nord, "and the republic elevated that cult into a secular religion, reverencing scientists as men of progress, raising statues to them and extolling their virtues to the young."[10] Scientific values filled the moral void left by the Catholic Church's diminishing role in French society. Furthermore, the first education minister of the Republic, Jules Simon (1814–96), consolidated the universities and lycées in order to mend regional divisions in France. Subsequently, Jules Ferry (1832–93) organized primary and secondary education to instill laïcité. Reforms in 1881 abolished fees and tuition for elementary schools; in 1882, enrollment was made compulsory. President Jules Grévy (1807–91) formally inscribed the separation of church and state into law on March 28, 1882, with a decree that barred religious instruction in state schools, leaving it up to families and private schools to teach religion.

Bergson belonged to the first generation of philosophy professors charged with the task of inculcating laïque and scientific values in the classroom. At the conclusion of his studies at the exclusive École normale supérieure in 1881, Bergson finished second in a class of eight who completed the state-administered agrégation.[11] Under the Third Republic, the exam greased the cogs of the educational bureaucracy. It served the function of channeling a corps of professors from Paris to the provinces. Moreover, passing the agrégation conferred an emblem of cultural prestige on young intellectuals; they belonged to the cult of "great men" who embodied the national genius.[12] As a result, philosophy looked quite different in the Third Republic. It was no longer a domain of independent thinkers. Maine de Biran was never a professor. Neither were eminent philosophers of the mid-nineteenth century such as Hippolyte Taine, Émile Littré, Charles Renouvier, and Antoine Cournot. These descendants of Enlightenment polymaths were replaced by academics who saw their trade as a profession.

After taking the examination, the next step for Bergson was to find a teaching position among the seventy-seven philosophy posts that were open in lycées.[13] He was first employed for a brief stint at Lycée d'Angers, a young women's school. He soon after moved to Clermont-Ferrand, where he developed the ideas for which he became famous: "At the École [normale], I immersed myself in mathematics and physics; I despised the rest, and made a 'bad lesson' on the agrégation in psychology, which I renounced. . . . It was in the contemplation of the province that I came around to completely changing my point of view: since, while reflecting there, I realized that mathematics could not explain time, as one perceives it within oneself; and so, everything that I had neglected up until that day as secondary became essential for me."[14]

Like most of his peers, Bergson set his sights on one of the sixteen philosophy chairs in the University of Paris system.[15] "Paris indeed remains the seat of all the exclusive resources enabling the exercise of intellectual power in the university: the scholarly reviews and societies, juries for recruitment and for the doctorate, national authorities of evaluation, collections of publications, and additional institutes for research."[16] Following two rejections for professorships at the Sorbonne, the École normale granted Bergson's wish in 1898. Soon after in 1900, Bergson leapt into the Collège de France, where he remained until his retirement in 1921.[17] By the age of forty, Bergson stood out among fellow philosophy professors for having rapidly ascended the educational ladder from the provincial lycées to France's premier teaching institution.[18]

Bergson stood out even more starkly, I want to suggest, for having taken advantage of the philosophy course. In the official curriculum released in 1874, psychology opened with a lesson distinguishing psychological and physiological facts. The lesson was not only the first in the class's history to recognize the cerebral bases of selfhood. It also established the precedent that psychology was to be presented under the rubric of *facts* amenable to experimental research. Soon after in 1880, the Ministry of Public Instruction again reformed the curriculum and presented psychology as "the science of *facts* and their *laws*," all the while acknowledging, "it is true that the order of facts, more than any other, provokes metaphysical curiosity."[19] Subsequent changes in 1885 and 1902 deepened the ministry's commitment to updating psychology. Although the lessons were mandatory, their content and organization were left to the professors' discretion. Profiting from an era of academic freedom, Bergson seized the opportunity. By the time he left lycée teaching, Bergson finished *Matter and Memory* (1896), the second of his two major books to tackle experimental psychology.

In the following sections, I trace the institutional contexts of psychology education from 1874 to 1902 and show how this backdrop explains the increasingly scientific character of French ideas of spirit. That transformation appears even more striking when we peer into life in the classroom. My aim is to foreground this intimate space where students engaged with philosophy professors by analyzing heretofore-neglected historical materials: course notes and philosophy textbooks. These materials open a revealing window onto the practice and experience of how neurophysiology was taught in the late nineteenth century. Notes and textbooks from Bergson's courses in particular lay bare the public mission of spiritualist materialism. They also explain the philosopher's meteoric rise to fame. Indeed, it is my contention that Bergson, so often construed as a singular thinker of the fin de siècle, climbed the academic hierarchy in large part because he effectively carried out the Republic's pedagogical campaign to bring together science and spirit in the classroom.

The Scientific Turn in Psychology Instruction

Lycées have existed in France since the Napoleonic reforms of the early nineteenth century. The system of secondary schools was instituted in order to reproduce cultural values for the aristocracy's children. For much of the nineteenth century, the majority of secondary students were not part of this public program. They attended ecclesiastic, or "free" schools. However, the Falloux law of 1850 decentralized education and charged each department with the obligation to ensure universal primary schooling and to expand secondary schooling. It was the responsibility of the towns to manage *collèges* (the French equivalent of junior high school). But lycées fell under the national purview. When the Third Republic took shape, it set about integrating all youth into the public schools and reorganizing philosophy instruction in the service of a nationalist project.

Following the Franco-Prussian War of 1870, French politicians and education reformers placed blame for the nation's defeat on retrograde scientific institutions. In the face of the technologically advanced German Empire, the Third Republic promoted scientific education as the key to surpassing the perpetual foe across the Rhine. In 1876, twenty-four university professors formed the Society for Higher Education in order to promote educational reforms and to collect information about foreign institutions. The organization's journal, the *International Education Bulletin,* dedicated the majority of its empirical studies to the German system. In France, education reformers straddled the competing imperatives of updating the curriculum with scientific research and preserving French philosophy's distinct spiritualist heritage. Even though, as the spiritualist philosopher Paul Janet reflected, "Germany has become our idol since it humiliated us,"[20] it was secondary education that distinguished the French. Philosophy instruction was not mandatory in the German gymnasium (the equivalent of high school). In France, collège included a section in practical moral education. But philosophy was unique to the final of year of lycée. It became the chief institutional conduit by which the Republic molded young citizens in accord with the scientific aspirations of the nation.

Philosophy instruction underwent four major reforms in 1874, 1880, 1885 and 1902 that progressively integrated experimental research into the psychology section. By 1907, the psychologist Alfred Binet had surveyed three hundred lycée philosophy professors to measure the impact of the reforms in the classroom. He concluded that, since 1874, "psychology has moved closer to biology and medicine: less introspection, in short, and more objectivity."[21] Among the 103 surveys returned, Binet noted a curious trend: "The influence of Mr. Bergson's ideas has, I believe, left the greatest mark. There are even four professors who adopt them without reserve while making them the heart of their teaching."[22] As one lycée professor wrote, Bergson's engagement with experimental psychology, "so lively and so rich, always seduces many students, at least those who understand something of it."[23]

Lycée students had the choice of pursuing one of two degree tracks in either classical or mathematical studies. With each reform, the importance of philosophy for those pursuing the latter diminished, while the content of philosophy in the classics degree became more scientific. This transformation was by no means uncontested. It reflected a confluence of intellectual and political power. Academics and professionals trained in universities made up a sizable portion of the government's deputies.[24] The Ministry of Public Instruction recruited philosophers into its ranks, including Ferdinand Buisson (1841–1932), the 1927 Nobel Peace Prize winner who served as director of primary education from 1879 to 1896; Élie Rabier (1846–1932), director of secondary education from 1889 to 1907; and Louis Liard (1846–1917), director of the universities from 1884 to 1902.[25]

Jules Simon first implemented his vision of secondary education as the head of the ministry from 1870 to 1873. His tenure stood out for its momentary stability during the tumultuous first decade of the Third Republic.

A former student of Cousin, Simon saw himself as carrying forth his master's legacy.[26] Simon envisioned philosophical instruction as the completion of scientific training. Left to itself, empirical knowledge abandoned students to educational pedantry. Drawing from Montaigne, Simon had developed a sardonic vision of schools deprived of philosophy's unifying force, "Just as birds sometimes go in quest of grain, and carry it in their beak without tasting it to give a beakful to their little ones, so our pedants go pillaging knowledge in books and lodging it only on the ends of their lips, in order merely to disgorge it and scatter it to the winds."[27] Scientific instruction required philosophy in order to be internalized by students. To pursue that goal, Simon replaced the imperial council of the university with a council of the ministry whose forty-eight members included numerous philosophers drawn from the university. The spiritualist philosophers Charles Jourdain, Michel Bréal, Ernest Bersot, Paul Janet, and Félix Ravaisson (who served as secretary) were "the friends," Simon wrote, "who met in my cabinet every Saturday and who, without an official title, by their friendship for the minister and above all for solid scholarship, work with me on all the reforms."[28]

It was on July 23, 1874, under the subsequent minister, Anselme Batbie, that the Ministry of Public Instruction released the first program reflecting Simon's vision. True to the Cousinian heritage, it opened with a brief introduction to the object of philosophy and its distinction from the sciences, followed by the first and most significant subject, psychology. But the 1874 reform also brought about a significant rupture: psychology ceased to be studied as a deductive inquiry and instead opened with a lesson on the nature of psychological *facts*. Demonstrative of the section's new empirical framework, the lesson distinguished psychological facts from physiological facts. The lesson acknowledged that metaphysics no longer laid claim to the only method appropriate to psychology. "The faculties of the soul," the second lesson in psychology, ceased to be understood as an absolute truth and was instead framed as a working hypothesis. Professors were thus expected to confront the biological approach to experimental psychology from the beginning of the class, effectively barring them from taking for granted the long-standing introspective approach of Cousin's spiritualist psychology. "It appears therefore that the general tone of this rubric was profoundly marked by the empiricism of a new discipline, psychology," Bruno Poucet writes, "a sign, along with the presence of political economy, of the defeat of the unity of the philosophy program, which no longer marked the triumph of spiritualism as the only point of view."[29]

Jules Ferry, a devotee of Comtian positivism, announced the new philosophy program on August 12, 1880.[30] The physiological orientation of psychology was reinforced. The curriculum continued to organize the subject around sensibility, intelligence, and the will; but these domains were no longer taught under the rubric of *faculties*. Sensibility, intelligence, and the will were instead presented within the experimental classification of distinct psychological *activities*, representative of the sensory-motor account of the self that had been adopted by psychologists. The lesson on liberty in the psychology section, for example, abandoned the Cartesian title "moral liberty or free will, its demonstration, and negation." The ministry refashioned the lesson into "the voluntary act," which presented experimental insights into instinctual behavior, reflexivity, and habituation alongside the metaphysical treatment of consciousness's immanent content. The reform established scientific and philosophical psychology as complementary fields.

Reviews were mixed among devotees of the spiritualist heritage. Alfred Fouillée celebrated the new program.[31] Others saw it as a direct assault on Cousin's faculty psychology. Francisque Bouillier (1813–99), a Cousinian loyalist, argued, "despite the authors of the program, the word 'faculty' is not more a part of our philosophical language than it is a part of our literary language. But the moment seems poorly chosen to ban the concept, while the reformers return more than ever to the honor of physiology, which they believe has definitively localized certain faculties in the brain."[32] The 1880 program included the new lesson "sleep, dreams, somnambulism, hallucinations, and madness," which introduced psychopathological research that fractured the unity of the Cousinian self. In its place, the psychology curriculum presented distinct modalities of psychic activity, buttressing a new spiritualism grounded on a dynamic, rather than substantial, conception of the self.

There was a growing gulf, however, between philosophical and scientific instruction. The ministry announced the new philosophy section for the mathematics degree on February 5, 1881. Rather than working from the philosophy program and trimming its lessons, as was the case in the 1874 reform, the ministry decreed an altogether distinct philosophy curriculum for mathematics students. It included lessons in the epistemological foundations of the sciences, which were cleanly distinguished from a section in morals. The separation reinforced the curriculum's nationalist orientation. The principles of just conduct and active citizenship were positioned as the terminus of empirical inquiry.

The 1880 reform was coupled with new academic freedom for instructors. An architect of the program, Henri Marion (1846–96), highlighted, "The order adopted in this program does not bind the liberty of the professor, provided that all the questions included are treated."[33] The reform opened the possibility for wide divergences among professors in both their pedagogical style and their treatment of the syllabus.[34] The newfound academic freedom had its skeptics. Paul Janet asked, "Do we believe that families can have faith in the University; isn't there a tacit accord which guarantees that the individual liberty of the professor does not pass certain limits and that he will not stray too far from those standard ideas on which society until now has rested?"[35] Such disputes over academic freedom fueled the emergence of pedagogy as a distinct domain of the human sciences. In 1883, Marion took over the first chair in the Science of Education at the Sorbonne. In his eyes, pedagogy was a branch of philosophy. It was accountable to principles transcending the whim of individual professors.[36]

Opening young women's secondary education was equally momentous. Thanks to the Republican deputy Camille Sée (1847–1919), the law of December 21, 1880, organized public collèges and lycées for girls.[37] The reform extended Ferry's secularizing imperative. In bringing young women under public tutelage, the state chipped away at religious schools' authority. Unlike secondary school for boys, there was no baccalauréat. Young women received a diploma of secondary studies (*d'études secondaires*), which was not a vocational degree. "For the great majority of those who voted for the law, women's access to schooling was not first of all designed for their personal blossoming, but rather for the stability and harmony of the household."[38] Hence Ferry's defense of the law in 1882, "Equality in education is the reconstituted unity of the family."[39] There was no Latin education or philosophy instruction, leaving young women without the materials deemed integral to the development of young men. During the third through fifth years, however, there was a morals class, which included Kant and a discussion of basic elements of psychology.

To ensure that local departments carried out the reforms, Ferry's ministry expanded the role of the inspectors general. Created in 1802 under Napoléon, the inspectors monitored all public schools. Under Ferry, two were assigned to primary schools, six to secondary schools (evenly shared between letters and mathematics), and eight to universities (three for letters; three for sciences; one for law; and one for medicine). Appearing in each classroom once a year, the inspectors evaluated the professors. The spiritualist philosopher Jules Lachelier, the professor of so many young philosophy students at the École normale supérieur, became one of the chief inspectors general for secondary education. He worked alongside the lycée philosophy professor François Evellin (1835–1910). Evellin and Lachelier saw to it that lessons were presented clearly and students were held accountable. In fact, the fate of all faculty members hinged on the inspectors' evaluations.

Lachelier served as an intermediary between the ministry and philosophy professors. When many expressed dissatisfaction that they had to learn new material, including brain anatomy and social science methods, it was up to Lachelier to propose a new curriculum.[40] What followed was a modified program on January 22, 1885.[41] Professors no longer taught political economy. It instead became the watered-down "relations between morals and political economy." Élie Rabier, who drafted the updated psychology section, pared away the metaphysical lessons "the idea of God" and "the idea of the external world." Lachelier renamed the "metaphysics and theodicy" section "metaphysics" and, against the scientific trend, added a lesson titled "providence and natural religion." Marion reorganized the logic section, further shedding its content. The final section, "history of philosophy," was replaced with "notions of the principal philosophical doctrines." It traced an intellectual lineage beginning with Socrates and, as if extending an olive branch to the old guard, concluded with Cousin.

An updated baccalauréat followed on January 28, 1890. For the degree in letters, philosophy constituted half the content. But it was reduced to 12 percent of the degree in mathematics, which was renamed "modern education" the next year. Worries had long brewed among philosophy professors about the fate of their discipline. "Philosophy retreats into letters and away from the sciences," wrote Émile Boutroux.[42] He inveighed against provincializing the philosophy course. Boutroux thought it instead "rests on science and on letters as two columns; and it collapses as soon as one is eliminated."[43] The widened gulf between the degrees in letters and mathematics had the effect of consigning philosophy instruction to a marginal role. The psychologist Théodule Ribot welcomed the declining significance of philosophy: "most [of the students] are put off by it; others intoxicate themselves in generalities and formulas that have no use."[44] Fouillée, however, went so far as to argue that all lycée professors ought to be trained in philosophy "so that all professors are, as much as possible, penetrated by the philosophical spirit."[45] In 1899, the former prime minister, Alexandre Ribot (1842–1923), commissioned a report including the input of Boutroux and Fouillée.[46]

What followed was the reform of 1902, which further fragmented philosophy instruction. Students now had the choice of pursuing four degrees instead of two.[47] Fouillée retorted, "So choose the easiest, the one that conforms the most to the taste you think you have."[48] Philosophy professors found it more difficult to recruit students into their classrooms. It had become possible for students pursuing degrees in medicine and law to study neither Greek nor Latin. In the eyes of many professors, the reform was the last blow to the linguistic backbone of philosophy instruction, shortsightedly cementing the vocational purpose of secondary education. One newspaper even portrayed the reform as France's "intellectual disarmament" before Germany.[49]

Released on May 28, 1902, the new philosophy program elevated the scientific approach by further trimming its metaphysical orientation.[50] "Our old philosophy must still be conserved," the director of the ministry announced, "but while *reducing its excessive dialectics* and *developing the scientific spirit*."[51] The metaphysics section was demoted to a mere three lessons instead of seven, the last of which was titled, "metaphysical relationship between science and ethics." The final section dedicated to the history of philosophy, over which Cousin's legacy still claimed a waning foothold, was eliminated. The same was the case for the philosophy program in the mathematic and medical tracks. No mention was made of the history of philosophy. Instead, the program was organized into two parts: elements of scientific philosophy and elements of ethical philosophy.

The 1902 program completely dropped the old spiritualist schema of sensibility, intelligence, and the will. It instead divided the psychology section into the dual categories of "intellectual life" and "affective and active life." The division cleanly separated physiology from metaphysics, paving the way for the concluding lesson reconciling "the physical and the spiritual." More importantly, the division lent expression to a defining commitment of spiritualist materialism—that metaphysical and experimental methods issue from complementary dimensions *within* the self. The reform brought the dual tensions propelling the educational upheavals since 1874 to a climax. More scientific than ever, on the one hand, yet distanced from scientific courses, on the other, the terminal philosophy course looked starkly different than its Cousinian antecedent. By the twentieth century, many believed the course had been dethroned from its exalted status atop secondary education.[52] Fouillée urged, "The philosophy class must remain *for everyone* a *terminal* class, the most obligatory of all, since it must raise *all* students above *all* of their particular studies."[53] His words rang hollow, the echoes, as it were, of a bygone era. The year 1902 brought a cessation to the chronic reforms to the philosophy course. The program remained until a contemporary version was released in 1960.

The period of educational reforms between 1874 and 1902 established the institutional conditions that drove philosophers to revamp the spiritualism of the past. Attuned to the new horizon opened by the brain sciences, yet simultaneously wary of their reductionist implications, education reformers lent support to a revivified strand of spiritualism, no longer grounded in faculty psychology, but nonetheless committed to the irreducibility of spiritual activity. A new generation of philosophy professors seized their lycée teaching posts as an opportunity to forgo the narrowly historical project of the bygone spiritualism and instead to mine incipient research on hallucinations and psychological abnormalities, the duration of attentive and perceptive psychic acts, and bourgeoning brain localization, including the cortical bases of memory and language (see table 1).

Confronting Neurophysiology in the Classroom

Lycée students pursuing a classical degree spent on average eight hours a week in the philosophy classroom. The typical day lasted fifteen hours, four of which were shared between two courses, while the rest were spent studying.[54] Most lycées were boarding schools where students lived throughout the year. Life inside left much to be desired. "Never-ending and obscure corridors, smoke-filled classrooms, bare and narrow lessons, the freezing atmosphere of dormitories, heady kitchen odors too close to the cafeteria, unclean lavatories, dusty courtyards, soulless parlors," so one former lycée administrator described conditions in the late nineteenth century.[55] The philosophy professor usually opened the class by dictating a summary of the day's lesson, which students copied in their notebooks. The French system promoted scrupulous note taking, to the point that students scrawled every "and," "but," and "however" of the professor's dictation. Students would rewrite their notes and submit them to the professor the next day. Following the lesson, students were welcome to pose questions. The professor would conclude by distributing essay prompts. "The professor expounds . . . and the student composes," as a typical dissertation manual presented the method. "The former is a kind of scientific work, the latter is a work of art."[56]

The inspectors general monitored philosophy professors closely. Whether the professor spoke clearly, held the students responsible, and used relevant examples were all criteria employed by inspectors when evaluating professors' performance. Philosophy professors were advised to begin the course with a philosophical lexicon: "the first duty of a professor is to create a rational vocabulary."[57] In his guide to pedagogy, Evellin outlined two teaching methods: the expository method, whereby professors read a lesson and answered questions from the previous day's, and the dialectical method, whereby professors posed a series of questions to lead students through the lesson ("cold calling" was standard practice). Evellin deemed the latter "the method the most appropriate for the formation of a youth, who, in order to become a resource for the country, must penetrate in all directions of thought, before we ourselves have even done so."[58]

In 1885, Evellin visited Bergson's class at Clermont-Ferrand and wrote

TABLE 1. Psychology sections of the philosophy curriculum

1874[a]	*1880*[b]	*1902*[c]
Psychological facts—consciousness—distinction between physiological and psychological facts	Object of psychology: distinctive character of facts it studies; degrees and the limits of consciousness	Specific character of psychological facts and consciousness
Faculties of the soul—sensation, intelligence, will	Distinction and relation of psychological and physiological facts	**Intellectual Life**
Sensation—sensations—sentiments instincts, penchants, passions	Sources of information in psychology: consciousness, language, history; utility of comparative psychology; experimentation in psychology; classification of psychological facts	The data of the understanding: Sensations; Images; Memory and association
Intelligence—external perception—internal perception—reason	Sensation: emotions (pleasure and pain); sensations and sentiments; inclinations and passions	Attention and reflection: The formation of abstract ideas; Judgment and reasoning
Ideas in general—classification of ideas—origin of ideas—different theories proposed on this question	Intelligence: acquisition, conservation, elaboration of knowledge	The creative activity of the mind
First principles, axioms, and ideas of reason	Acquisition: data of consciousness and of the senses	Signs: relations between language and thought
Intellectual operations—memory—association of ideas—imagination	Conservation and combination: memory, association of ideas, imagination	Principles of reason: development and role
Attention, abstraction—comparison—generalization	Elaboration: formation of abstract and general ideas; judgment, reasoning	Formation of the idea of the body and perception of the exterior world
Judgment—reasoning	Directing principles of knowledge: data of reason; can they be explained by experience, association of ideas, or heredity?	**Affective and Active Life**
Signs and language—relations of language with thought	Results of intellectual activity: idea of the self, idea of the external world, idea of God	Pleasure and pain; Emotions and passions; Sympathy and imitation
Will. Instinct. Habit	Aesthetic notions: beauty, art, principles and conditions of the fine arts, expression, imitation, fiction, and the ideal	Inclinations; Instincts; Habit
Moral liberty or free will. Demonstration of freedom—the principal systems that deny freedom	Will—analysis of the voluntary act: freedom	Will and character; Freedom
Harmony of the faculties of the soul—unity of the principle of these faculties—human personality	Various modes of psychological activity: instinct, voluntary activity, habit	**Conclusion**
Spirituality of the soul [*l'âme*]. Distinction between body and soul; their union; laws of this union—different systems that deny the distinction between body and soul	Manifestations of psychological life: signs and language	The physical and the spiritual; Psychological automatism; Personality: the idea of the self
	Relations of body and mind [moral]—sleep, dreams, somnambulism, hallucinations, madness	
	Elements of comparative psychology	

[a] *Bulletin Administratif du ministère de l'instruction publique* 17 (1874): 490–93.
[b] Henri Marion, "Le nouveau programme," *Revue philosophique de la France et de l'étranger* 10 (1880): 427.
[c] *Bulletin Administratif du ministère de l'instruction publique* 71 (1902): 760–62.

a positive report.[59] Bergson recounted, "It was at Clermont that I made my most essential discoveries. But the inspectors general, [François] Evellin and [Jules] Lachelier, who inspected me there, forcefully told me: you have to go to Paris and take your place."[60] Shortly after publishing *Time and Free Will* in 1889, Bergson left for Paris to teach at the prestigious Lycée Henri IV. Bergson's evaluations helped to expedite his ascent up the academic hierarchy. In 1894, Lachelier returned to observe Bergson as a professor in Paris. The inspector praised the "rigor of his method," returning the next year to write another glowing report: "complete clarity compatible with his depth, and only here and there are some things a bit artificial for his thought and a bit thin on development."[61] Soon after in 1896, Bergson was selected to teach courses in Greek and Latin philosophy as a docent at the Collège de France, where he took a full professorship in 1900.

Bergson had a natural talent for oratory. But he achieved notoriety as a thinker for having deftly handled the psychology curriculum. Teaching motivated Bergson to read scientific articles voraciously. And he treated his lycée course as an experimental forum for bringing neurophysiology into dialogue with spiritualist philosophy.

An essay prompt scrawled in the notebook of one of Bergson's students in an 1893 course reveals the imbrication of scientific and metaphysical concepts in the classroom. Bergson had students respond to the question "What does the philosophical spirit consist of? Determine the nature of the philosophical spirit by opposition with the scientific spirit and common sense."[62] "Common sense" carried the particular meaning of a phenomenon's appearance, which traditionally functioned as the starting point of dialectic reasoning in French philosophy instruction. Taking the concept of external perception as his example, the student compared, first, the utility of perception for survival according to common sense; second, perception as an optical function; and third, the knowledge that perception gleans from reality according to epistemology. The student concluded, "there is therefore a real opposition between the philosophical spirit and the scientific spirit."[63] The dual structure of the outline reflected Bergson's method of setting scientific and metaphysical concepts in a mutual dialogue.

Contrasting Bergson's performance in the classroom with that of his spiritualist contemporary Jules Lagneau (1851–94) throws into high relief the pedagogical divergence between scientifically revamped concepts of spirit and their eclectic predecessors. A fellow graduate of the École normale who passed the agrégation in 1875, six year before Bergson, Lagneau pursued the same career trajectory. He hopped from lycées in Sens (1876–78), Saint-Quentin (1879–80), and Nancy (1880–86). Lagneau made it to Paris in 1886, taking a professorship at Lycée Michelet, where he remained until his premature death at the age of forty-three. In contrast to Bergson's magisterial production, Lagneau managed to publish only a few review articles during his lifetime.[64] He achieved belated fame as a lycée professor. Thanks to his student Émile Chartier (1868–1951), the future philosophy professor who adopted Alain as his nom de plume, Lagneau's notes enjoyed posthumous publication. But Lagneau hardly invoked scientific examples in the classroom. As a result, he failed to earn the respect of the inspectors general, which accounts for his failure to ascend to a Parisian university.[65]

Lagneau dedicated the bulk of his philosophy course to psychology. As one student commented, "During the major part of the academic year, he only treated the introduction to philosophy and psychology extensively. All the rest was taught briefly or passed on by means of copied texts."[66] Lagneau's approach to psychology, however, was markedly less engaged with neurophysiology than Bergson's. Lagneau swiftly handled the introductory psychology lesson on the distinction between psychological and physiological facts.[67] A student in an 1886 course jotted in his notes, "It is known that in order to conceive of certain thoughts, we have to make use of a part of the brain. . . . One might be tempted to substitute psychological knowledge for physiological knowledge, which arrives at tangible results." Like Bergson, Lagneau believed that there is a strict separation between the two. "Physiological science can only go back to conditions of conditions: it would not know how to seek the reason for facts, but only the conditions in which they are produced."[68] But rather than engage with neurophysiology, Lagneau dismissed its relevance. In a similar fashion, Lagneau moved quickly past required topics, "comparative or descriptive psychology," "physiological or explanatory psychology," and even "psychophysics," under the lesson "the objective form and experimental method in psychology."[69] Lagneau made zero references to experimental psychologists. It was thus unsurprising to find that inspectors as early as 1879 wrote in their reports: "Course too metaphysical."[70]

When Lachelier came to evaluate Lagneau's class in 1887, he wrote a positive review: "His class is one of the strongest that I have seen this year; not only do the students respond well and voluntarily, but they also handle the subjects with ease and clarity in their essays."[71] But when another inspector, Élie Rabier, visited two years later, the report was hardly as warm: "Without a doubt, no one can teach contrary to his doctrine, and unfortunately yours, I do think, is among the most difficult to teach because of the originality of your point of view, which is completely metaphysical."[72] Rabier advised Lagneau to temper his abstract lessons with at least three scientific examples per week, "so that students could study sensations as ordinary facts."[73]

Bergson, by contrast, tackled the same lesson on the distinction between physiological and psychological facts by forcefully articulating the stakes of brain research. He cited Wilhelm Wundt and Paul Broca before placing these experimentalists in the historical lineage of cerebral localization, beginning with the early nineteenth-century phrenologists Franz Gall and Johann Spurzheim. "Our moral life consists of science, art, and religion, but we cannot at all see how nerves cells, if they existed alone, could coordinate themselves in a manner to bring about these great thoughts and beautiful feelings."[74] In fact, Bergson's words were prescient. The term "neuron" first appeared in 1891.[75] In the beginning of the twentieth century, scientists came to acknowledge that the nervous system is composed of independent cells, each of which has an arboreal structure (as the anatomy resembles a tree): a body (soma), an extending axon, and several dendrites branching out.[76] Electrical impulses called *synapses* were shown to pass between dendrites and axons, facilitating the exchange of neural information.[77] The studies of the Spanish neurologist Santiago Ramón y Cajal (1852–1934) were foundational. He refuted the reticular theory of nervous composition, which had held that nerve fibers form a continuous web.[78] Using a silver chromate staining technique, he died thin slices of the cerebellum, optic nerve, and spinal cords of birds. From these colored stains, Ramón y Cajal conjectured, "Each nerve cell is a totally autonomous physiological canton."[79] Bergson was lecturing about the brain on the precipice of a paradigm shift. A new way of thinking about cerebral matter was afoot across Europe. And in French classrooms, a new of way thinking about spirit had blossomed.

In the lesson on psychophysics, Bergson laid out Fechner's theory as well as that of like-minded psychophysicists Joseph Delboeuf and Ewald Hering (1834–1918). "The experiments and Fechner's law are very debatable, even for those who admit the possibility of calculating sensation. We could even go further: we can wonder if a similar law would not be vicious in its very principle, for what does it mean for sensations to double, triple, or quadruple another?"[80] Bergson entered into and elaborated on psychophysics' research program in order to arrive at the thesis of *Time and Free Will*: the immediate data of consciousness are distinctly qualitative.[81]

Psychology was central to both Bergson's and Lagneau's teaching. I want to suggest that the difference turned on these professors' respective engagements with neurophysiology. Whereas Lagneau took metaphysics as psychology's point of departure, Bergson took it as the field's point of arrival. A young Alain wrote in his notes for Lagneau's lesson "psychology and metaphysics," "the true science of the spirit is not psychology, but metaphysics," since metaphysics reveals the principle of the unity of the self on which thinking depends.[82] Conversely, Bergson oriented his psychology lessons around research in clinical and experimental psychology in order to demonstrate that it led to conceptual problems that would generate new metaphysical concepts. Between the two professors lay divergent approaches to spirit. Lagneau's rigid commitments gained little purchase in an intellectual climate marked by the rapid influx of physiological methods into psychology. Bergson, however, opened the study of selfhood onto the sciences.

Bergson and Lagneau used the same psychology textbook, Élie Rabier's *Lessons of Philosophy*.[83] Originally published in 1884, the manual was published in twelve subsequent editions until its last in 1912.[84] A heavy tome of 676 pages, Rabier's textbook was dedicated solely to psychology. It was the author's belief that "philosophy collaborates with scientific studies, in the sense that it must first of all better understand science, and appreciate it all the more."[85] *Lessons* was one of the most widely used textbooks of the period.[86] In fact, it was the same textbook that the young Marcel Proust (1871–1922) used as a lycée student.[87] Perhaps its success was due to the fact that the author served in prominent educational positions, first as inspector general before becoming the director of secondary instruction in 1889. It is likely that Lagneau used the textbook because he received a free copy.[88] But Alain suggested that it was merely out of reverence to the inspector general that Lagneau mentioned it at all during his lectures: "Rabier was the 'pedant' . . . kept in hand during his inspection."[89]

It is clear from the notes taken by Bergson's students that he regularly incorporated and even took issue with Rabier's textbook. In its lesson on consciousness, the textbook disputed the notion of the unconscious. For most experimental psychologists of the late nineteenth century, the unconscious referred to a domain of automatic physiological processes taking place outside of consciousness. "The hypothesis of the unconscious is useless, since all the services that one claims of the *absolute* unconscious can be easily asked of a *relative* unconscious (made of small perceptions), since nothing stops us from admitting that the gradations of consciousness go until infinity."[90] Notes from Bergson's course at Clermont-Ferrand in 1887 indicate that he remained faithful to Rabier's claim.[91] After he transitioned to Paris, however, Bergson took advantage of his elevated teaching position and leveled a critique. Adding an additional lesson on the problem of unconscious sensations, Bergson asked, "How can we explain them? A first solution would consist of purely and simply denying the possibility of unconscious psychic facts. That is where several contemporary psychologists stop, including Rabier. Their argument can be summed up thus: a psychological fact is by its very definition a conscious fact."[92] Bergson presented the work of Wundt and Herbert Spencer, as well as materialist philosophers such as Hippolyte Taine and Rudolf Lötze, who countenanced sensations governed by mechanical laws operating outside of consciousness. Bergson disagreed. He nevertheless cautioned students that it would be "too easy to decide the issue quickly" and instructed them to appreciate the problem at stake: "Without a doubt, the perfectly psychological state is a conscious state, but this property of psychological states, as important as it may be, is not the only one. Even though there are states that might not resemble the properties of conscious, psychological facts, they are infinitely more conscious than physiological or physical facts."[93] Rather than set the physiological domain of the unconscious in opposition to the spiritual domain of consciousness, Bergson invited students to investigate their points of contact. He showed that spirit limns embodied experience.

It should not be surprising that physiological notions of the unconscious played such an integral role in Bergson's teaching. They figured prominently in *Matter and Memory*, in which Bergson argued that memories could not be localized in the brain's neural tissue. Memories do not reside outside of consciousness but instead inhere in the deep strata of consciousness. His approach hewed to the method that he had promoted in the classroom: "Without teaching philosophy to children," he reflected, "how can we bring them to pose for themselves, even in a very vague form, some of the problems to which philosophy seeks the solution? . . . These problems emerge naturally in biology, physics, and even in mathematics."[94]

Science and Metaphysics in the Culture of Textbooks

The new era of academic freedom under the Third Republic galvanized the market for textbooks. Professors had their choice.[95] Philosophy rode the wave as new publishing houses sprung up in Paris. Between 1874 and 1879, there were twenty new philosophy textbooks; nine between 1880 and 1884; and fourteen more following the 1885 reform.[96] With the relative calm after the reform of 1902, a mere six new philosophy textbooks appeared during the first decade of the twentieth century.[97] Most were written by philosophy professors who sought to achieve momentary fame by publishing their course notes, often times in lieu of more significant scholarship.[98]

Philosophy textbooks shared a generic form. They opened with an introduction in which the author clarified recent reforms to the program that prompted a new edition. When he was honest, the author would also make his philosophical commitments explicit. This was a characteristic feature of the textbooks used during the first half of the nineteenth century. "The doctrines of the manual are the pure and strict spiritualist doctrines," a typical

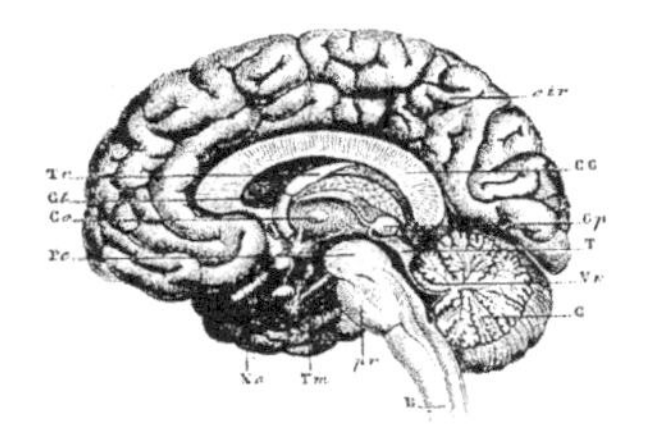

FIGURE 2. Textbook diagram of vertical section of brain
Source: Paul Janet, *Traité élémentaire de philosophie à l'usage des classes* (Paris: Delgrave, 1879), 21.

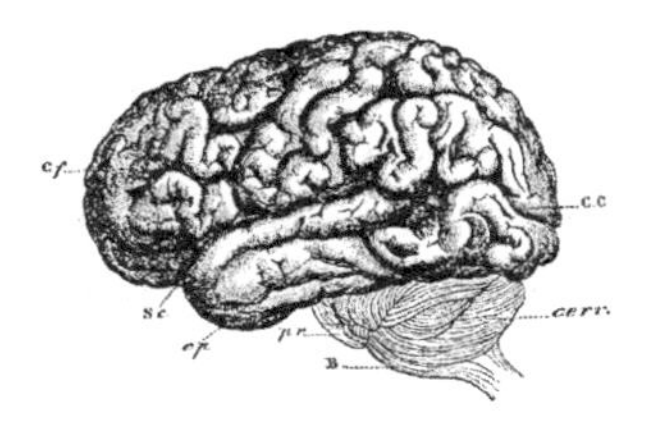

FIGURE 3. Textbook diagram of lateral section of brain
Source: Paul Janet, *Traité élémentaire de philosophie*, 22.

introduction read, "which the University, under the impulse and leadership of an illustrious philosopher [read: Cousin], arduously sets about propagating."[99] With their increasingly scientific temperament under the Third Republic, fewer textbooks conveyed such idolatry. A list of the lessons contained in the program followed the introduction. The bulk of the textbook featured the contents of the lessons, with italicized headings indicating the most important paragraphs. The greatest differences turned on the principles, philosophers, and examples that their authors chose to include.

Notable among the textbooks that Bergson used in his classrooms was Paul Janet's *Elementary Philosophy Treatise*.[100] The textbook was trailblazing for having been the first to feature diagrams of the brain's dual hemispheres and to acknowledge the cerebral bases of the self (see figures 2 and 3). There were also diagrams of the ear, eye, nervous system, and spinal cord. Conceding that contemporary advances in psychology left the old spiritualism outdated, Janet opened the textbook with a significant neurophysiology lesson. He grounded his approach in the claim that "all philosophy must depart from what really exists."[101] Bergson used Janet's popular textbook both for its novel analysis of neurophysiology and for its trenchant refutation of those who derived reductionist implications from the brain sciences: "By leaving out a discussion of the body and the role it plays in our life, we leave a dangerous weapon in the hands of materialism; for this part of our being, which, set up and displayed in its truth, cannot at all jeopardize what is higher in us."[102] The brain sciences were thus mobilized in support of the revamped spiritualist curriculum.

Janet saw his manual as a strategic line of attack in the face of the "philosophical crisis," which he had announced in a book by the same name in 1865.[103] In that book, he refuted the reductionist claims of scientific materialism. With the educational reforms, Janet saw an opportunity to appropriate that very materialism. The physiology lesson consisted of a thorough description of the human body, the organs and their nutritional functions, and the anatomy and functions of the nervous system, including an extensive discussion of the reflex arc. Subsequent chapters of *Elementary Philosophy Treatise* advanced higher in the hierarchy of psychological complexity, from affective phenomena, sensations, and memory, to recent psychopathological discoveries on sleep, dreams, and madness.

Janet divided his textbook into two classes of psychology, "on the one hand those aspects which immediately pertain to the body, and which we share in common with animals, and on the other hand those aspects which raise us higher and belong only to man."[104] It was along these lines that Janet distinguished physiological and metaphysical methods respectively; and it was central to philosophy's mission to demonstrate that the former, pursued to the limits of brain studies' explanatory power, intractably lead to problems of selfhood, notably the relation between spirit and matter, that only spiritualist concepts could address. This epistemological division underwrote a pedagogical strategy designed to preserve the autonomy of metaphysical and moral principles thanks to the negative support of physiological facts.

The appearance of Janet's *Elementary Philosophy Treatise* marked a transformative moment for the spiritualist stalwart. "Such an innovation, were it introduced some years before, would have been powerfully audacious, an individual revolt against academic traditions," Serge Nicolas notes. "But [Janet's] attempt in 1879 indicated, on the one hand, that the academy began to open up to the new psychology . . . and on the other hand, that Janet himself had undergone a kind of conversion."[105] The scientific turn in lycée psychology instruction took place in parallel with the conceptual arc of Janet's oeuvre, from his trenchant anti-materialism to a surprising affinity for the brain sciences. Janet had ridden Cousin's coattails into one of the three philosophy chairs of the Sorbonne in 1862. He initially defended his master's eclectic spiritualism against the mounting flood of brain research in Paris. In a series of polemic articles published in 1867 as *The Brain and Thought*, Janet argued that neurology, especially as it had been advanced by the French physiologists Jean-Pierre Flourens (1794–1867) and Louis-Françisque Lélut (1804–77), remained tethered to the phrenological picture of mind-brain relations, according to which mental activities issue from determinate cerebral locales. Janet argued that psychology instead depends on the transcendental principle of conscious unity, which could not be deduced from biological facts about the brain. He admitted that the brain does constitute a facet of the self, but only its material condition. "Thought results from the conflict established between cerebral forces entrusted with external actions and the internal force or thinking force, the principle of unity, lone possible center of the conscious individual."[106] Janet's antipathy toward physiological methods gradually subsided. As a gatekeeper of young philosophy professors' advancement, Janet came to open the doors of the Paris philosophy faculty to neurophysiological methods in psychology. By 1886, he proved especially congenial to the young Théodule Ribot, whose candidacy Janet supported for the first chair in experimental psychology of the University of Paris system.[107] Neurophysiological methods, as Janet came to acknowledge, need not threaten spiritualists' commitment to the autonomy of the will.

Janet's *Elementary Philosophy Treatise* set the precedent for future philosophy textbooks such as Abel Rey's *Elementary Psychology and Philosophy Lessons* (1903), which included even more extensive diagrams of the brain and nervous system (see figures 4 and 5). Rey, a philosophy professor and historian of science at the Université de Dijon, launched his career by publishing the most advanced psychology textbook following the 1902 reform to the program. His textbook featured diagrams of the sympathetic and autonomic nervous systems. Sensory-motor functions were represented in tables localizing nervous centers in the brain's grey matter, including the centers of linguistic memory (divided into the motor images of writing, vocal motor images, vocal auditory images, and visual images of words) as well as the centers of sight, taste, and smell. Although hardly a spiritualist partisan, Rey nonetheless employed Janet's epistemological division of labor between physiological and metaphysical psychology in order to "avoid distorting the minds of young students, by *carefully distinguishing what fits scientific study and what is the simple object of philosophical reflection*."[108] The division reflected Rey's commitment to psychophysical parallelism, the doctrine that nervous transmission and conscious activity are two aspects of the same psychic phenomena. But the division equally served philosophy's public mission to demonstrate the autonomy of metaphysical and moral principles.

These textbooks contrasted sharply with bygone generations of text-

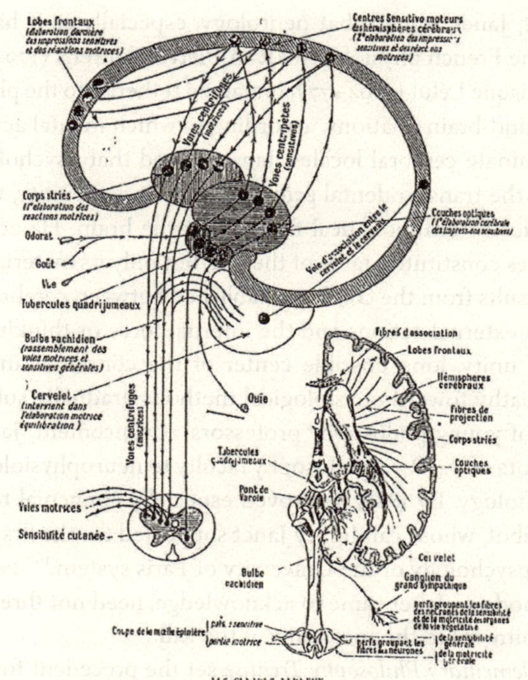

FIGURE 4. Textbook diagram of the central nervous system
Source: Abel Rey, *Leçons élémentaires de psychologie et de philosophie* (Paris: Édouard Cornély, 1903), 29.

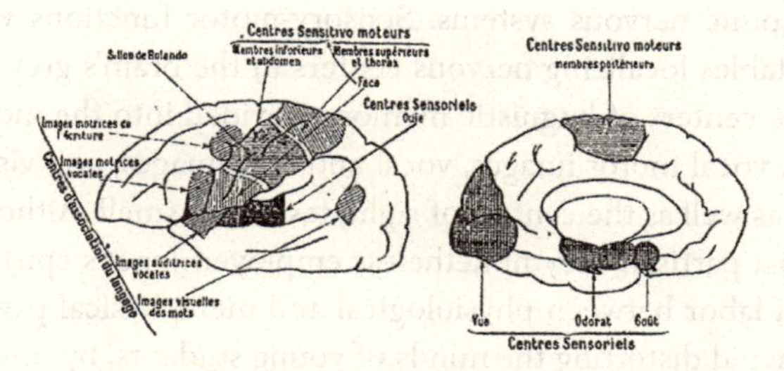

FIGURE 5. Textbook diagram of cerebral localizations
Source: Abel Rey, *Leçons élémentaires de psychologie*, 144.

books that advanced the old spiritualism. Charles Jourdain's (1817–86) *Notions of Philosophy* was the longest-lasting philosophy textbook to survive the education reforms of the Third Republic. Originally published in 1847, the textbook was published in eighteen editions until its last in 1888. Jourdain wrote the first version on the basis of the notes he had prepared while teaching the philosophy course at Collège Stanislas. He added more with subsequent editions. Since the collège was private, affording him a wider range of curricular choices, Jourdain freely promoted Christian teachings. His subsequent work for the Ministry of Public Instruction provided an official platform from which Jourdain disseminated the manual for public schools as well.

The first edition was originally published as *Questions of Philosophy* with a mere 127 pages.[109] Philosophy instruction was briefly suppressed under the Second Empire, during which time Jourdain renamed his textbook *Notions of Logic*. Despite the change in name, Jourdain highlighted the central importance of metaphysics and ethics: "On the questions that interest [students'] morality and happiness, like free will, duty, and fate, God prepared for them the interior lights that not even the laziness of the spirit and misbehavior of the will can successfully extinguish."[110] Following the reinstatement of philosophy in 1864, Jourdain changed the title to *Notions of Philosophy*. He could again champion Cousinian principles.

Jourdain saw the Republic's reforms as an affront to Cousin's eclectic spiritualism. In subsequent editions of the textbook, Jourdain impugned Jules Ferry's measures. "We continue to offer to the youth positive affirmations of the soul and God, while casting aside those equivocal conclusions that betray the master and implant dangerous uncertainties in the student's thought."[111] In Jourdain's eyes, Ferry had inscribed moral relativism into the curriculum. It was a criticism that Jourdain pursued in an 1880 screed, *School without God*, in which he lambasted the laïque turn in education.[112] He believed that religious instruction was not only necessary to establish the foundations of metaphysics and ethics. In his textbook, Jourdain argued that it also elevated philosophy above the life sciences. In contrast to the materialist moment in spiritualism, which had become preoccupied by the problems broached within the sciences, Jourdain toed the Cousinian line, arguing that philosophy ought to establish the first principles on which the sciences rely. For example, in the history of philosophy section, Jourdain pilloried Herbert Spencer and the advocates of association psychology, who had made students "relive this false and pernicious doctrine taught today in their work, with a talent worthy of a better cause."[113] In the final, 1888 edition of *Notions of Philosophy*, Jourdain clung to Cousin's bygone spiritualism by claiming, "whatever the new evolutions of philosophical systems, a point remains constant in our eyes: they will efface from man's thought neither the certainty of the existence of the soul, a soul endowed with the liberty and responsibility for his acts, nor the certainty of the existence of God, a personal, creating and rewarding God."[114] Jourdain's long-enduring philosophy textbook was a paradigmatic example of the fading yet tenacious influence of the old spiritualism over the classroom.

George Fonsegrive's (1852–1917) *Elements of Philosophy* offers a striking contrast. Written following the 1885 reform on the basis of notes composed while a professor at Lycée Buffon, Fonsegrive's textbook was dedicated exclusively to psychology.[115] Each lesson began with a summary of key concepts, followed by a résumé. At the conclusion of each lesson, Fonsegrive included a list of suggested readings, frequently citing recently published literature, and even experimental research.[116] *Elements of Philosophy* was organized into two sections: affective psychology and reflexive psychology. Affective psychology, he claimed, studied unconscious phenomena not attributable to the self. The section offered an extensive survey of research pertaining to the nervous system and perception, as well as to the structure of the brain. Reflexive psychology, which Fonsegrive presented as "far more interesting than the former," pertained to conscious phenomena.[117] The section featured the conceptual problems of consciousness left unresolved by experimentation: for example, the composition of images, the structure of conscious activity, attention and its relations to muscular states, maladies of memory, and the formation of sensorial atlases. In accord with the shared intellectual archive of spiritualist materialism, Fonsegrive cited Maine de Biran, and not Cousin, as the motivation for the textbook's organization.[118]

Fonsegrive was a Catholic indebted to the spiritualist tradition. Yet his courses at the École normale under Émile Boutroux propelled Fonsegrive to seek a rapprochement between religious and scientific thought. In his dissertation, *Essay on Free Will* (1887), Fonsegrive argued that determinism and free will are compatible. The laws of the natural sciences are the products of human creativity, a claim that echoed Boutroux's thesis in *The Contingency of the Laws of Nature*.[119] Fonsegrive's textbook conveyed his philosophical commitment. "It does not seem impossible to adopt these results, all the while remaining faithful to the doctrinal traditions precious to the University."[120] By placing science in dialogue with the history of Western philosophy, *Elements of Philosophy*, according to one reviewer, left "little in the way of bringing together the theories of modern psychology, as they are represented by the authors that everyone knows."[121] After the textbooks proved to be a success, Fonsegrive dedicated his career to bringing psychological advancements to bear on his Catholic faith.[122]

⁂

As psychology and neurology progressively illuminated the material foundations of selfhood in late nineteenth-century Europe, a lycée philosophy curriculum emerged under the Third Republic that promoted these emergent sciences and simultaneously steered their public reception. The educational reforms issued between 1874 and 1902 endowed generations of professors no longer teaching under Cousin's academic authority with the responsibility of redirecting the materialist implications of brain inquiry toward a renewed focus on the corporeal points of connection between spirit and matter. The evolving philosophy program, and the scientific turn in its psychology curriculum, precipitated a significant rupture in France's spiritualist heritage. The textbooks and course notes from the most impressive teacher of the period, Henri Bergson, reveal an educational campaign that sought to stake out the newfound pertinence of the nation's spiritualist tradition on the basis of the human and life sciences.

Lessons in experimental psychology and neurophysiology bolstered the metaphysical and moral content of the philosophy curriculum. The organization of the curriculum around the relations between psychological and physiological facts reflected a disciplinary organization that set humanistic inquiry atop the sciences. The organization also proved socially efficacious. The psychology section ensured that the young brain sciences would not entail a reductive materialism; they would instead mold an active citizenry grounded in the nation's distinct intellectual heritage and sustained by Republicans' imperative to disseminate emergent scientific knowledge. These educational institutions supported conversations between scientific and philosophical knowledge. Moreover, the materialist turn in the spiritualist curriculum generated new avenues for the study of the self. In the words of one textbook, "Metaphysics begins there where science and experiments can no longer say anything."[123]

CHAPTER 4

Locating Selfhood in the Brain

In late nineteenth-century France, mapping the self along the topography of the brain came to preoccupy the public imagination. Endeavors to draw up atlases of the organ's functions were abounding. This was the aim of cerebral localization: to uncover the neural regions governing distinct mental activities. That a visual guide could reveal the material coordinates of selfhood was of a piece with the rational and secular imperatives of the French Third Republic. Localization research eschewed the long-standing Christian belief that the brain was a homogeneous organ, the seat of a soul whose unity had been imparted by a single act of God's grace. A pervasive faith in scientific progress motivated neurologists and psychologists to chart the inner contours of spirit along the heterogeneous convolutions of the cerebral cortex.

Memory was especially prominent among the spiritual powers thought to be "found" in the brain. In his 1881 book *Diseases of Memory* (*Les maladies de la mémoire*), the psychologist Théodule Ribot fostered a widespread fascination with the abnormal.[1] He showcased reports of expressive aphasia. Patients with the disorder were unable to articulate words aloud. When presented with a fork, for instance, they could neither name the object nor utter the word "fork."[2] Ribot surmised that damage to neural tissues had destroyed their ability to vocalize words. He followed the pathological method, examining the symptoms of forgetting in order to deduce the cerebral architecture of remembering. His conclusion was sweeping: "Memory is not, following the vague expression of everyday language, 'in the soul': it is fixed in its place of birth, in a part of the nervous system."[3]

Pathological investigation propelled advancements in cerebral localization (see figure 6). But there was a problem. Although aphasiacs could not enunciate the word "fork," they could recall the idea, nodding their head when asked if the object was a fork. It appeared the memory was not lost.

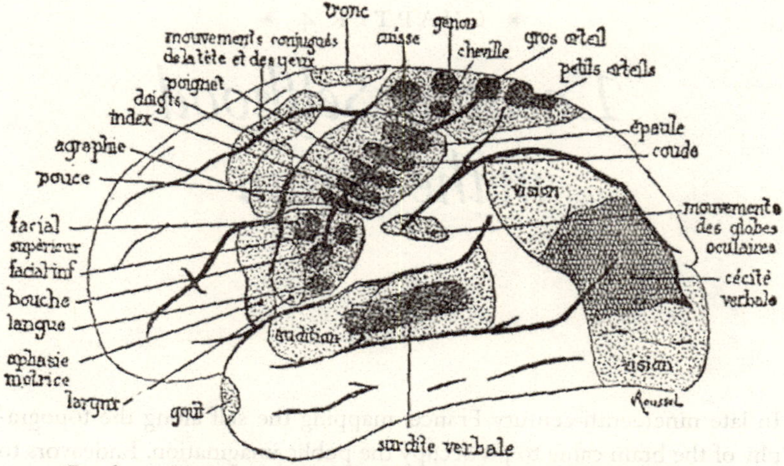

FIGURE 6. Diagram of localizations of sensory and motor regions in the left cerebral hemisphere
Note: In the diagram, the majority of localizations correspond to the face and limbs; these were identified along the motor cortex.
Source: Maurice Georges Debove and Charles Achard, *Manuel de médecine*, vol. 4, *Maladies du système nerveux* (Paris: Rueff, 1894), 260.

A spiritual remainder endured despite neural damage. Where would this memory—unmoored from its cerebral underpinning—fit on a map of the brain?

Whether aphasiacs' memory was material or immaterial cut to the heart of the fraught relations among science, selfhood, and society in the fin de siècle. At stake was not only the fraught connections between spirit and matter, but also the future of epistemological authority in Republican France. In their efforts to promote a reductionist ontology that would explain everything (including memory) in terms of cerebral matter, scientific thinkers such as Ribot insisted that neurology and psychology—rather than philosophy or theology—spoke for the rational nature of selfhood. They believed that the case of aphasia was a reminder of the inchoate state of brain research. The patient match of laboratory inquiry, aided by continued government investment, would soon enough integrate the totality of memory into a complete understanding of the cortex.

In the eyes of Henri Bergson, science was moving in a different direction. The convergence of cerebral localization and the psychopathology of memory disorders, instead of having laid metaphysical problems to rest, had actually generated new dualisms. In his 1896 treatise *Matter and Mem-*

ory (*Matière et mémoire*), Bergson argued that aphasia revealed a duality within memory: between mental representations and embodied practices. The former are interwoven with a rich inner world comprising flashes from the past, episodes of dreams, and psychic events—all of which Bergson called "image-memories." Scientists had failed to locate them in the brain because the organ's physiology is disposed not to represent ideas in the mind but to facilitate sensory-motor action in the material world. For Bergson, "the brain is no more than a kind of central telephonic exchange: its office is to allow communication, or to delay it."[4] In a telephone exchange, workers connected callers' lines at a switchboard where electrical impulses traveled from long distances (see figure 7). The figure illustrated Bergson's belief that aphasia incapacitated only the *transmission* of memories. The neural mechanism for expressing the word "fork" had been short-circuited; this motor-memory ceased to partner with the image-memory. Consequently, the disorder lay bare the riven structure of memory. Far from justifying a reductive materialism that localized the past in the brain, Bergson found aphasia to support a spiritualist materialism premised on the corporeal points of contact through which memory entered into and interacted with cerebral matter.

FIGURE 7. Central Telephone Bureau in Paris
Source: "Les demoiselles du téléphone: Aspect d'un bureau téléphonique parisien," Bibliothèque nationale de France, Département des Estampes et de la Photographie (1904), http://expositions.bnf.fr/proust/grand/193.htm.

In this chapter, I explore the incredible resurgence of spiritualist concepts amid the rise of localization research. In the final decades of the nineteenth century, the aporias of brain inquiry overflowed the working categories of clinical and experimental study, producing a conceptual surplus that called for a metaphysical resolution.

That philosophers engaged so closely with the work of neurologists and psychologists may seem striking given our firm disciplinary borders today. The 1870s through the 1890s was an era of incredible advancement in neurophysiology. New understandings of the brain migrated across disciplines. The anatomical organization of the brain into four lobes had only recently taken hold. Their functional relations came to be analyzed with ever greater precision.[5] Localization research upended the principle of cerebral equipotentiality, according to which all parts of the cortex were equally responsible for volitional activity. The principle had persisted in the French neurological community since the physiologist Jean-Pierre Flourens had claimed that damage to any region of the brain impaired the entirety of coordination. More importantly, a new concept of matter came into being. The heterogeneous regions of the cortex entailed a conception of somatic material far different than the homogeneous atoms that had been the building blocks of classical physics' mechanistic world. In light of these developments, it may be tempting to look back on ideas of spirit as a recalcitrant intellectual bulwark against scientific progress, especially given Bergson's argument that "the cerebral process answers only to a very small part of memory."[6] Indeed, the claim that memory is not confined to the brain has garnered a fair share of criticism.[7] But I see the significance of Bergson's and other spiritualists' thought to hinge more so on the underlying conviction that brain science was not confined to science. Animating the materialist moment in ideas of spirit was a growing recognition that knowledge about the organ is entangled with the self's cultural milieu.

It was not accidental that Bergson figured the brain as a telephone exchange. In an era of extraordinary industrial production, technological metaphors were commonplace. They resonated across the environmental, social, and intellectual landscape of Europe. Thinking about the "wiring" of the nervous system took place against the backdrop of urban skylines filled with webs of telephone cables. Copper wires, conductors, and poles saturated the visual field. By the 1880s, telephone cables crossed national boundaries, brought distant interlocutors into contact, and forged connections among disparate communities. The technology emerged out of the telegraphic networks that linked post offices, weather observatories, government agencies, and banks across Europe over the course of the nineteenth century. It was often taken for granted, in the words of one physiologist, that nervous conduction functioned "as the telegraphic wire that transmits impulsions."[8] The metaphor came into scientific and popular parlance alongside other industrial innovations such as photography, radio, and phonographs. The rhetoric fused machinery and corporeality into a system of "organic physics," leaving an indelible trace on understandings of the nervous system.[9] Even today, it is not uncommon to talk about our "hard wired" brain (even though the organ is soft and wet).

It was in this rhetorical ecosystem, I want to suggest, that spiritualist thinkers clarified and contested the conceptual surplus of brain science. At issue were not *mere* metaphors. Figuration played a structuring role in concepts of the brain. Today, researchers rightly eschew the argot of wiring along with the notions that brains "process" information or that neurons "light up."[10] As much as the neurosciences disavow such figures now, it is not because they are alien to the field. Technological figuration emerged within neurophysiology during the nineteenth century. In France, the industrial climate was the mise-en-scène in which neurologists, physiologists, and psychologists thought, wrote, and argued about neural matter. But there was no predetermined script. Although metaphors were everywhere, they did not entail a single concept of memory, selfhood, or the nervous system. Metaphors set the stage on which argument took place.

Technological vocabularies were conceptual tools deployed to collapse spirit and matter and also to cleave them apart. On the one hand, spiritualists mobilized metaphors as persuasive means to elucidate the dualities of selfhood. Memories were figured as immaterial "messages" propagated via the brain's material wiring. On the other hand, the metaphor served reductionist thinkers who figured memory as an electrical impulse that automatically traveled along the lines of nervous circuits. Laboratory inquiry and machinic imagery shared cultural resonances, making possible a domain of contestation in which diverse intellectual communities came together.

Metaphors were elastic. As I argue in the following sections, spiritualist thinkers exploited this elasticity. For them, technological figures did not apply exclusively to the body. Nor did the interlocking discourses of nervous vibration and electrical propagation only engender an atmosphere of energy, shock, and speed—tropes that historians have taken to be characteristic of modernity.[11] Industrial rhetoric also lent expression to the conservation of the past, the solitude of recollection, and the experience of time within the inner life of the self.

A growing body of work in history, cultural theory, science studies, and anthropology has brought attention to the formative contributions of technological figures to the human sciences.[12] In particular, Laura Otis has shed light on the conjoined organic and mechanical worlds in which electricity furnished a cultural feedback loop, as neurologists and industrialists reinforced the affinity between nerves and wires.[13] Friedrich Kittler has shown how the telegraph, typewriter, and telephone affected the ways that people corresponded and understood their bodies.[14] The idea that metaphors shape scientific theories is by no means new. The spiritualist philosopher Jean-Marie Guyau was keenly aware, writing in 1880, that "in the as yet imperfect state of science, metaphors are absolutely necessary: before we know we have to start by figuring."[15] The languages of culture and the cortex were porous. This is not to suggest that the human sciences simply invoked rhetorical devices as techniques of vulgarization.[16] My argument is that technological figures offered a frame that thinkers deployed in order to bring neurophysiology to bear on broader issues of subjectivity.[17] Electricity, communications, and machinery provided a shared idiom that brought disputes over the commitments and consequences of brain research to the forefront of fin de siècle thought.

Sensation, Motion, and the Spiritual Brain

Brain-mapping projects originated as part of broader inquiries into the self. Pierre Paul Broca's (1824–80) work was trailblazing. The medic founded the French Society of Anthropology in 1859.[18] A chief preoccupation was craniometry: the practice of measuring the volume of skulls.[19] Broca would pour lead shot into a hollowed cranium and then drain the substance into a graduated cylinder. The data were compiled in a cephalic index (the maximum width of the head multiplied by 100 and divided by its maximum length), which Broca used to organize brain sizes into racial hierarchies. Unsurprisingly, the typical French cranium was deemed intellectually superior, thus fabricating an "objective" rationale for the country's mission to civilize colonized peoples of North Africa in the late nineteenth century.[20] Ideology and science were interwoven.

Broca's racist approach to brain science cannot be conveniently separated from the results that have proved to be foundational. At the Bicêtre Hospital outside Paris, Broca worked with a certain Mr. Leborgne, or "Tan" as he was known. The patient was given the name after convulsively repeating the phrase "tan, tan" whenever questioned by clinicians. He suffered from what was called "expressive aphemia"—the inability to articulate words. Broca noticed that Tan nevertheless comprehended what was asked of him. "How many years have you been in Bicêtre?" Broca would ask. "He opened his hand four times and then added one finger. That made 21 years, the correct answer."[21] After his death, Broca dissected Tan's brain. Since the right side of the patient's body had been paralyzed, Broca presumed that he would find damage in the left hemisphere. He relied on the still nascent belief that the brain's dual hemispheres were functionally contralateral.[22] Once the grey mush was set on a table, there it was: a lesion to the inferior frontal gyrus (the third convolution of the cortex). Broca inferred that the damage was the cause of Tan's symptoms. In a series of monumental reports, Broca announced that he had uncovered the seat of language, what came to be known as "Broca's area."

The discovery breathed new life into cerebral localization. Identifying mental activities in the brain was already thought to be possible. As I explored in the first chapter, the Viennese anatomist Franz Gall arrived in France in 1807 and went on to lecture widely on cranioscopy. He exhibited cranial charts, which identified the protrusions on the skull corresponding to psychological traits—a project that was of great interest to the early spiritualist Maine de Biran. Although cranioscopy (or phrenology, as it was subsequently named) was discredited as a dubious—and eugenic—approach to the brain, Broca's pathological investigations corroborated Gall's underlying vision. His method, however, was flipped on its head. Whereas Gall devised a table of mental faculties and proceeded to search for them on the surface of the skull, Broca took neurophysiology as his starting point. Associating neural damage with the symptoms of memory loss laid the clinical groundwork for cerebral localization into the twentieth century.

Broca catapulted a metaphysical problem to the center of the scientific community. Just what had been located in the brain? His explanation of Tan's lesion teetered on a fine line between spiritual and material notions of language: "What is lost . . . is not the faculty of language, it is not the memory of the words nor is it the action of the nerves and of muscles of phonation and articulation."[23] Instead, it was "the *faculty* of articulated language."[24] In a subsequent report, Broca characterized the functions of the inferior frontal gyrus thus: "the part of the encephalon linked to intellectual phenomena and of which the cerebral organs are, as it were, just the agents."[25] Broca's claim to have discovered the material "agents" of a mental power emboldened ideas of spirit in France. For the medic Pierre Foissac (1801–86), the discovery "brings us to consider Broca's circumvolution as a psychomotor organ of speech only."[26] In demarcating the motor function of the neural region, Foissac meant to distinguish it from the immaterial will. "If we intentionally insist on the question of cerebral localizations, it is because, far from leading to materialism, there has never been a question of anatomy and physiology that has furnished such convincing, such obvious proofs of the spirituality of the soul."[27] Localization research thus appeared to uncover the cerebral locales that served as the instruments of volition.

Subsequent experiments reinforced the idea that spiritual life intervened in the brain. Working on a dressing table in their Berlin home, Gustav Fritsch (1838–1927) and Eduard Hitzig (1838–1907) applied an electric current to dogs' brains. The scientists used a trephine to drill holes in the skulls. With the exposed brain section, Fritsch and Hitzig stuck pins in the regions where minimal stimulation produced maximal movement in the animals' limbs. Each trial allowed them to access the skull with greater precision. In a 1870 study, they produced a brain map of the motor locales associated with muscular movement. "From the sum of all of our experiments," Fritsch and Hitzig claimed, "certainly some psychological functions and perhaps all of them, *in order to enter matter or originate from it* need certain circumscribed centers of the cortex."[28] The report was integral to the budding idea that the human brain, like a dog's, has a motor cortex. Crucially, the authors suggested that the localizations indicated where the will came into contact with the nervous system.

The mysterious relations between spiritual activity and cerebral motility inflected further localization experiments. In Britain, the neurologist David Ferrier (1843–1928) also used electrical stimulation. He irritated monkeys' cortices to provoke muscular convulsions in the animals. His work nearly sent him to jail after Ferrier was indicted for violating the British "Cruelty to Animals Act" in 1881.[29] The stimulated monkey brains provided evidence for his chart of fifteen areas, as he characterized them, for "the localisation of cerebral function, not merely as regards motion, but also as regards sensation and the other faculties of mind."[30] That Ferrier claimed to uncover mental powers alongside sensory and motor locales entrenched the metaphysical dualities within localization research.

Others sought to eliminate the possibility of dualism by reformulating cerebral matter as exclusively sensory-motor. Sensation and motion served as the basic neural constituents for the decomposition and analysis of higher mental activities. The categories were central to the work of Carl Wernicke (1848–1905), who in 1874 identified the center of language comprehension in the superior temporal gyrus, what came to be known as "Wernicke's area." He followed Broca's pathological method. Wernicke examined the postmortem brains of patients afflicted with receptive aphasia. They were unable to convey or understand meaning. Although the patients were able to utter words, their speech was reduced to fluid nonsense. In locating the sensory complement of Broca's area, Wernicke brought neurologists to see that the latter actually governed motor functions. Contrary to Broca's belief that he had discovered the seat of a language *faculty*, Wernicke argued, "nothing could be worse for the study of aphasia than to consider the intellectual disturbance associated with aphasia as an essential part of the disease picture."[31] In his eyes, a map of the brain served to orient only the neural activity of the self.

The brain's organization was called into question.[32] For Wernicke, the organ comprised not discrete locales, but overlapping neural connections. He developed the theory during his medical studies in Vienna as an assistant to Theodor Meynert (1833–92). The neurologist proposed an interconnected picture of the brain. A "projection system" of neural elements connected the brain with the external world. And an "association system" integrated cerebral tracts running between clusters of memory images (also called "memory pictures" or "impulses").[33] According to Wernicke, Broca's area was an interface situated across the systems. Damage to the area impeded the flow of neural information. Vocal memory-images were part of a sensory-motor network that also included clusters of auditory memory-images around Wernicke's area as well visual memory-images in the angular gyrus (part of the parietal lobe). The German neurologist Ludwig Lichtheim (1845–1928) brought together Broca's and Wernicke's studies into a compatible model by formulating the neural pathways between the respective areas.[34] After examining conduction aphasia (a disease that left patients unable to repeat words despite vocal and auditory comprehension), Lichtheim claimed that motor and auditory image-centers connected to a higher center of semantic memory images. The varied etiologies of aphasia thus came to be understood as memory disorders nested in the crossroads of a sensory-motor web.

Was the brain a mosaic of separate regions? Or was it a constellation of neural interconnections? Disputes over the composition of cerebral matter were inseparable from problems of spirit.[35] It was in this conflictual intellectual context that diverse thinkers grappled with the broader implications of localization research for the nature of selfhood.

Nervous Wiring in the Built Environment

Contestations over the spiritual and material dimensions of the self reached a crescendo in a technological atmosphere filled with telegraph wires. By the mid-1800s, thousands of miles of above-ground wires ran along Europe's railroads. Wernicke invoked the metaphor to characterize the brain's association network. He likened "the whole process to the forwarding of a telegram," in which the region of motor speech was analogous to "the telegraph station that transmits the telegram."[36] The metaphor was especially potent at a time when the electrical nature of nervous conduction was still a relatively new discovery. The German scientist Hermann von Helmholtz had originally proposed the analogy between nerve fibers and telegraph cables.[37] They functioned as the material conduit of nervous action. For the German neurologist Emil du Bois-Reymond, "between the two apparatuses, the nervous system and the electric telegraph, there is . . . more than similarity; it is a kinship between the two, an agreement not merely of the effects, but also perhaps of the causes."[38] The figure applied not only to nervous transmission. The telegraph was also a *graphic* device meant for inscription.[39] It was illustrative of the translation of nerve impulses into linguistic signs—a rhetorical liaison between motion and meaning. But anatomical accuracy was left wanting. Nerves were slower and not as smooth as actual wires. Du Bois-Reymond cautioned, "unlike those wires, [nerves] do not, once cut, recover their conducting power when their ends are caused to meet again."[40] Metaphors did not establish an identity between nerves and telegraph wires. Nor were sensory organs and muscular activity equated with transmitters and receivers. The rhetoric allowed scientists to pick out salient aspects of the structure, function, and composition of human biology.

In particular, "wiring" served as a synecdoche for the entire relay of afferent and efferent nervous action from the body's periphery, to the spine, and back out. Connections across the nervous system were imagined in continuity with the electric lines that proliferated across Europe, linking cities in a transnational communications nexus. One author declared, "The network of cables that surround France also surround the entire world."[41] In fact, the first transatlantic telegraph line was established in 1858. It stretched between Newfoundland and Ireland, making it possible to send electric signals across continents. The wired brain signified a node in an intercontinental network across which messages relayed at unheralded speed. The physiologist Charles Richet claimed: "Just as the telegraphic network that unites all the cities of Europe makes it so an event that happens in one of these cities reverberates almost immediately in all the others, so does the excitation of a single cell communicate to all the other cells of the body through the nervous network."[42] Nerves were not confined to nations. As electromagnetic and electrochemical telegraphs fueled the expansion of cabled communication across the globe, the "circuitry" of the nervous system served to harmonize the brain, body, and self.

The meaning of communications rhetoric was not always straightforward. Copper wire seemed appropriate to describe the material composition of neural fibers. Could the figure also apply to the metaphysically fraught problem of spirit's connection to sensory-motor action?

In France, numerous thinkers deployed telegraphic figures in order to highlight the dualities of matter and spirit. In their eyes, mental and cerebral activity were independent processes. For Victor Egger, the latter "is the function *of* the organ, it is the organ in function, the organ in movement; it is the organ and something else."[43] In the mind, however, "the function is sui generis, apart, without relation to the organ, heterogeneous to the organ and all movement; it is a world apart, the unextended world."[44] The telegraph offered an illustration of how the two dimensions of selfhood were irreducible to each other. Alfred Fouillée suggested that they operated like rival impulses flowing through the circuitry of the nervous system. For Fouillée, nerves played a significant role in the body as the corporeal *conduit* by which physiological forces entered into confrontation with mental forces. "The cells of the cortex are mainly trophic and act like combustible materials placed along a line of gunpowder. Be that as it may, one could not figure the brain as a powerless telegraph worker who would passively read dispatches in one direction without being able to transmit them consciously in the other direction. Everything that happens in the nervous system must have its counterpart in the mental, both the output of the current as well as the input."[45]

In Fouillée's eyes, mental and cerebral activity traversed the nervous systems in a dynamic tension. He reminded readers that telegraphic relays were not unidirectional. Signals were sent back and forth simultaneously in an electrical circuit. Moreover, Fouillée brought attention to the fact that telegrams—the series of dots and dashes created by making and breaking the circuit—had to be interpreted by an operator. The process was not automatic. His argument hewed to that of Julius Robert Mayer (1814–78), the German physicist who claimed, "We know that no telegraphic dispatch can take place without the concomitant production of a chemical action. But what the telegraph says, in other words the content of the dispatch, cannot be considered in any fashion as a function of electrochemical action."[46] He highlighted the epistemic gap between meaning and motion. "The brain is only the instrument, it is not itself spirit."[47] Telegraphic signals, like sensory-motor action, were translated from Morse code thanks to a thinking subject.

Telegraphic figures equally buttressed reductionist interpretations of brain mapping. Neural connections were imagined as "*communications* permitting a great variety of reflex phenomena."[48] In his neurology handbook *The Brain and Its Functions* (*Le cerveau et ses fonctions*) (1876), Jules Bernard Luys (1828–97) figured sensory-motor transmission "as a series of electrical wires, stretched between two stations, following two principal directions."[49] The metaphor supported Luys's crude materialism. He argued that mental life emerged out of "organic phosphorescence," the property by which nerves persist in a "vibratory state." It also conveyed the malleability of neural connections. According to the English neurologist Henry Charlton Bastian (1837–1915), "There is no difficulty in supposing that many nervous currents may pass though one of these compound nerve fibres, just as many electric currents might pass simultaneously through a single telegraphic or telephonic wire."[50] As for the architect of the association paradigm, Theodor Meynert shied away from technological tropes. He preferred zoological rhetoric. Meynert figured the brain as jellyfish: "Just as medusae stretch out their feelers into the world and take possession of their prey through tentacles, so this composite protoplasmic being, the cortex, possesses centripetally conducting extensions, the sensory fibers of the nervous system, which we may consider its feelers, and motor fibers, which are its tentacles."[51] To his mind, the brain was a vital organ and therefore more supple than the mechanical operations of telegraphy.

In Paris, communications networks were somewhat unique. The city had cables installed underground, hiding them from public view. Beyond the metropolis, aerial wires ran along train tracks, connecting the cities of France with the rest of Europe. In the urban landscape, however, wiring came to be conspicuously absent. Cables were visible nowhere yet extended everywhere. The telegraphic network was a subterranean—and subconscious—facet of experience.

The specter of wiring haunted scientific literature. Lexicons of oscillation, undulation, and vibration construed nervous action in a manner that was suggestive of wired communication. For instance, it was common "to suppose that the nervous centers of sensation feel particular shocks in relation to the physical qualities of waves transmitted by nervous fibers."[52] There were also affinities with the elasticity of the ether, the medium through which light and sound were presumed to travel. The rhetoric was polyvalent. According to Hippolyte Taine, "meaning and imagination, sensation, perception—in short, thought—is only a vibration of cerebral cells, a dance of molecules."[53] Vibratory propagation was often a thinly veiled allusion to wired operations. It served to tether the self to the brain. Nervous contagion, the phenomenon of subliminal mimicry (notably in psychiatric studies of hypnosis), was conceived as a property of "sound waves and light waves that are like the result, the continuation of the cerebral movement and nervous movement that gave birth to them."[54] For the sociologist Gabriel Tarde (1843–1904), such metaphors were symptomatic of a materialist zeitgeist. "Everything in the physical world is effectively reduced to . . . undulation; everything takes on more and more an essentially undulatory character."[55] Tarde cautioned scientists to recognize where they relied on metaphors. He found it "arbitrary as well as purely verbal, the comparison of the nervous network to the telegraph or telephone."[56] Charles Richet did not see a problem. "There is probably a nervous current, or rather a wave, a vibration. It also seems useful to call the modifications in excited nerves a *nervous vibration.* This denomination doesn't suppose any preconceived idea and it is convenient for exhibiting facts."[57] In Richet's eyes, technological tropes were pragmatic tools.

Not all mapping projects relied on wiring metaphors. Beginning in 1883, Jean-Martin Charcot (1825–93) examined aphasia at the Salpêtrière Hospital in Paris. Before the neurologist became celebrated for showcasing hysteric women at weekly public demonstrations, Charcot exhibited aphasiacs singing and playing musical instruments before audiences. He synthesized his findings into a bell-shaped diagram composed of five memory centers: two sensory centers each for listening and reading, two motor centers for speaking and writing, and a fifth intellectual center.[58] One of the attendees, a young Sigmund Freud (1856–1939), went on to identify the mechanisms of speech in a cortical interface extending from visual and auditory centers on one end to motor regions on the other. Before he embarked on his mature psychoanalytic work, Freud wrote *On Aphasia* (*Zur Auffassung der Aphasien*) (1891), in which he argued that damage to the entire web, and not a single locale, resulted in memory loss.

Some psychologists eschewed the telegraph metaphor. Théodule Ribot thought it was inappropriate to explain memory.[59] He believed that wiring failed to do justice to the complex entwinement of psychology and biology. Ribot emphasized that the brain is "not like a central point from which everything radiates and to which everything leads (as the pineal gland of Descartes) but is like a prodigiously and inextricably tangled lattice in which histology, anatomy, and physiology constantly get lost."[60] Nonetheless, metaphors were inescapable. Ribot invoked parliamentary figures to make sense of the brain: "Step by step, from delegations to delegations, visceral life would find its final representation there."[61] For Ribot, the brain was a distributed system without any single center of memory, a government whose leader had been deposed. The physiologist Léonce Manouvrier (1850–1927) invoked similar imagery in order to emphasize the hierarchical structure of the nervous system: "The government receives contributions from all the federated states (nervous currents) as well as the information that it centralizes. It receives at the same time external information (external senses). Thus richer and better informed than each, and more disinterested as well, it distributes the funds received to the particular interests and general prosperity in a manner even more exact and completely informed."[62] The figure reinforced visions of the French "body politic."[63] The Third Republic was envisioned as the apex of a social organism whose legitimacy depended on the authority of science.

The fraught relations between mental activity and sensory-motor action resonated across diverse intellectual communities in large part thanks to a rhetorical ecosystem that enveloped pathological inquiry and localization research. Brain studies were mobilized to advance past intellectual traditions as well as future scientific aspirations. The meaning of technological lexicons was hardly stable. They ramified contestations over the problems that emerged from laboratory and clinical studies of memory's neural bases. Instead of having laid the dualities of spirit and matter to rest, metaphors established a conceptual proving ground for thinkers to bring brain science to bear on competing accounts of selfhood.

Time, Space, and the Structure of the Past

Advancements in cerebral localization left little doubt that the brain governed significant functions of memory. Still unanswered were vital questions about the nature of those functions. Growing acceptance of the sensory-motor paradigm made it clear that neural centers facilitated recollection and retention insofar as these were embodied activities. The corporeal modalities underlying memory functions (e.g., writing, listening, and speaking) were being identified with greater precision. Troubling, however, were memories that depended on neither sensation nor motion, that were not instantiated in the present but embedded in the bowels of the past. How could the transient passage of time be represented in the spatial configurations of the cortex?

According to a central tenet of French spiritualism, memory eclipsed the brain. For many thinkers, psychopathological and localizationist methods had revealed varieties of aphasia that brushed only the surface of a vast temporal reservoir at the core of selfhood. According to Fouillée, "time alone is the essential form of memory, and memory, ultimately being the consciousness of appetite, effort, and motor action, is as fundamental as life itself."[64] He and other thinkers took it on themselves to articulate the temporal conditions that made possible the brain's memory functions. Memory, they reasoned, was more than a collection of discrete episodes; it was also an energetic force animating our subjectivity. Bergson made the point evocatively: "The process of localizing a recollection in the past . . . cannot at all consist, as has been said, in plunging into the mass of our memories, as into a bag, to draw out memories, closer and closer to each other, between which the memory to be localized may find its place."[65] The picture was suited to the spatial configuration of the cortex, in which discrete episodes of the past might be deposited. Although we recall individual memories in the present, Bergson believed that memory was not conserved in the past as a collection of particularities. Between the depths of time and the surface of the cerebrum lay a metaphysical abyss that neurologists and psychologists struggled to explain: namely, the spiritual power by which the past pressed on the present.

Spatial metaphors were widespread in the science of memory. Ribot conceptualized the recollection of past episodes as "a retrogressive march, which, starting from the present, traverses a more or less extended series of terms."[66] Traversing the entire chain of memories, as if reading every dossier of an archive, would be exhausting. So we make use of "landmarks" (i.e., significant life moments) that transport us back in time. The spatial rhetoric was legible. Ribot remarked "how this mechanism resembles that of localization in space. Here, also, we have our reference points, abridged methods, and well-known distances which we employ as units of measurement."[67] Ribot's theory built on the prevailing understanding of memories' organization in the brain, according to which the underlying associations were embedded in neural tissue. The English physiologist Henry Maudsley (1835–1918) had shown that since all memories involved embodied movements, they leave "motor residue" in the brain. "Such are the physical bases of memory," Ribot inferred.[68] Motor residues facilitated the smooth execution of bodily habits, what he called "organic associations." With repetition over time, they formed a neural support network for higher "psychological associations." As Ribot saw it, the associative structure of memory depended on the brain's neurophysiological architecture. He stressed, "it is impossible to determine where memory—whether psychical [i.e., mental] or organic—ends."[69] The strata of memory corresponded to the degrees of complexity in the brain's webs of motor residues.

In aphasia, the bonds between the organic and the psychological broke down. Linguistic signs progressively dissociated from ideas in the order that they had originally associated. This came to be known as "Ribot's law." "The march of amnesia" followed the same trajectory: patients lost the ability to recall words, spoken and written, beginning with names and continuing through nouns, adjectives, and finally verbs. Language regression pursued "the path of least of resistance."[70] Since verbs are repeated the most often, they forged the strongest associations, making them the last to erode. Crucial to Ribot's law was the work of the English physiologist John Hughlings Jackson (1835–1911), who showed that cerebral disorders led to psychic regression. Nervous functions were lost in the inverse order that they had developed. For example, Ribot noted that the dissolution of higher rational language often involved an acute sense of lower emotional language. Similarly, aphasiacs who lost recent memories reported glimmers of a forgotten past. Underlying the structure of memory disorders was Ribot's foundational commitment: "memory is, *per se*, a biological fact—by accident, a psychological fact."[71]

A curious line of argument among spiritualists was that the associationist account of aphasia neglected the role of volition in language. Fouillée and Bergson duly revised Ribot's law of memory loss along these lines. They argued that verbs were resistant to aphasia, not because they were reinforced by stronger neural associations, but because verbs expressed effortful aspects of memory. According to Fouillée, "verbs, both passive and active, which survive the longest, are the immediate expression of emotions and actions."[72] Verbs were conceptualized as a component of language allied to the body's motility. For these thinkers, the brain's motor action was not an automatic process. It also involved what Bergson called an "effort of the mind."[73] To recall a memory, we strain our mental energy, furrow our brows, and concentrate on a personal past. It followed that "verbs, in general, which essentially express *imitable actions*, are precisely the words that a bodily effort might enable us to recapture when the function of language has all but escaped us."[74] Wholly different were nouns and adjectives; these were abstract components of language. Bergson reimagined aphasia on account of the temporal organization of memory: riven between action in the present and a contemplative retreat into the past. "The fact can be explained if we admit that memories need, for their actualization, a motor ally, and that they require for their recall a kind of mental attitude which must itself be engrafted upon the attitude of the body."[75] The upshot was a bit peculiar. It seems that Fouillée and Bergson conflated the meaning of words with their syntactic function. Moreover, the argument would likely not obtain in languages other than French. English has numerous proper names that are also embodied actions: Chase, Carrie, Tripp, Wade, Cruz, or Mark.

More pressing were the lacunae of psychopathology, which prompted a reimagining of the temporal structure of the past. Case studies revealed that aphasia affected only motor-memories. Patients afflicted with the disorder receded into the their private thoughts. "The memory of vocal and written signs which survives in the intelligent subject of aphasia represents what is called the 'internal voice' [*parole intérieure*], that minimum of determination without which the mind would be on the way to dementia."[76] Ribot was convinced that scientific progress would eventually locate the neural underpinning of the phenomenon. "The relation between sign and idea, a simple fact in subjective psychology, becomes in positive psychology a complex problem which can only be solved with a further development of our knowledge of anatomy and physiology."[77] The internal voice opened a glimpse into a vast expanse of memory beyond the limits of motor action in the present.

Victor Egger's *The Internal Monologue* (*La parole intérieure*) (1881) was an early attempt to delineate the temporal and the spatial. Egger argued that aphasiacs' image-memories, untethered as they were from motor-memories, constituted a limit case for cerebral localization. His goal was to distinguish image-memories and motor-memories from the first-person perspective of recollection. Pointing to the subtitle, *An Essay in Descriptive Psychology*, one reviewer noted, "The only method he employs is observation by consciousness or rather by memory."[78] The thesis bolstered spiritualism's claim to be a science of the inner self. Jules Lachelier praised "the way that you start between the data of consciousness that we take for ourselves and those that we let go into the external world."[79] From this inner perspective, private thoughts appeared as distinctly spiritual data of consciousness. Whether it unfolded unknowingly or preoccupied our attention, our internal monologue constituted an immaterial dimension of language persisting in excess of motor vocalization. Egger hedged much of his argument against the English psychologist Alexander Bain, who believed that buccal activity (mouth movement) accompanied intellectual states. Egger pointed to cases in which an internal monologue took on the voice of others, a phenomenon heightened in hallucinations. One felt aware of the sounds of speech even though they found no corporeal expression. Egger's account contradicted those of neurologists. For Wernicke, the internal monologue constituted a "word conception";[80] the neurologist Joseph Dejerine (1849–1917) called it a "word notion" (*notion du mot*).[81] Both aligned with the faint rhythm produced by the body's motor language capacities. For Egger, the inner monologue confounded cerebral localization. He believed the deeper the descent into memory, the more freely that private thought detached from sensory-motor functions.

The distinction within memory—between motor speech and the inner monologue—corresponded to the spatial and temporal dimensions of selfhood. In parsimonious terms, Egger posited, "in affirming space, we affirm the non-self [*non-moi*]; in affirming duration [*durée*], we affirm the self; the unextended that endures is the self."[82] To think in spatial terms was to adopt the perspective of third-person observation and thereby seek out the cerebral locales of memory. Such a perspective entailed a departure from immediate experience. In a letter, Egger clarified: "The theory comes together by an effort to analytically bring together the self and duration, the non-self and extension: the extended milieu of phenomena is *constructed* by spirit; the enduring milieu of phenomena is *given* without rupture to consciousness."[83] The latter corresponded to the domain of memory, which

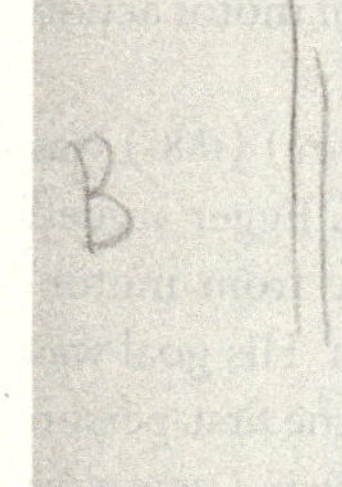

Selon nous, l'amnésie des signes est surtout une maladie de la *mémoire motrice ;* c'est là ce qui lui donne son caractère propre, ce qui fait qu'elle s'offre sous un aspect nouveau. Mais que faut-il donc entendre par « mémoire motrice », expression qui au premier abord peut surprendre ? C'est une question si peu étudiée par les psychologues qu'il est difficile d'en parler clairement en passant et qu'il est impossible de la traiter ici tout au long.

FIGURE 8. Marginal note in Bergson's personal copy of Ribot's *Les maladies de la mémoire*
Source: Fonds Henri Bergson, Bibliothèque littéraire Jacques Doucet, BGN—II 40, 148.

was interwoven with the processual life of inner subjectivity. "The self is what flows, what passes or has past, but which, once past, often becomes present again."[84]

Clinical cases revealed a fundamental duality between time and space. This was the conceptual scaffolding on which subsequent spiritualist thinkers intervened in debates over psychopathology. The rupture between aphasiacs' image-memories and motor-memories was pivotal. Language was shown to be one form of embodied memory. As Fouillée argued, "each organ is a memory; the eye is a memory of luminous waves and the ear is a memory of sonorous waves."[85] Image-memories, however, flow in the stream of time (what might be called the "subconscious," the term in use during the era). Pathological cases were a point of departure to imagine the immaterial structure of temporality within the self.

Bergson's archive is telling. He spent five years studying the literature on aphasia. "As for cerebral localization," Bergson recounted, "I never doubted it for a single moment, since I envisioned only . . . what had been rigorously proven for the functions of speech. The question posed itself for me in another form. What mattered was to determine the exact *signification* of the facts of localization."[86] We find Bergson's interpretive practice on display in his personal copy of Ribot's *Diseases of Memory*. Bergson scribbled a giant "B," likely signifying *bien* (good) next to the passage: "amnesia of signs is above all a disease of *motor memory*; that is what gives it its distinctive character" (see figure 8).[87] The emphasis with which he made the marginal note suggests that Bergson took Ribot to have delimited the brain's motor functions, thus leaving open the possibility of a spiritual realm of memory. What followed for Bergson was a simple yet precise duality within memory: the nature of the past is to be no longer present. Motor-memories act; image-memories are dead.[88] Contrary to association psychology, which held there to be a difference in degree between the two varieties of memory, Bergson interpreted clinical studies to imply a difference in kind.

The duality within memory mapped onto two basic operations. The *recollection* of memory—as Fouillée succinctly put it, "this judgment that the image of a thing is a simple image"—was for spiritualist thinkers not at all tantamount to the *conservation* of memory.[89] Jean-Marie Guyau elegantly framed the difference. He argued that psychologists too often considered "the brain in a state of rest" and the memories within it "as fixed, *clichés*."[90] Memories recalled in the present were qualitatively different than those conserved in the past. As Bergson presented it, recollection served perception (what he called the organism's "attention to life"). To fixate on image-memories was to erect a mental buffer from practical demands of the *hic et nunc*. "To call up the past in the form of an image we must be able to withdraw ourselves from the action of the moment, we must have the power to value the useless, we must have the will to dream."[91] The mind thus plunges beyond the limits of the present, probing the depths of the past conserved.

Although they seemed to function as distinct modules of the mind, recollection and conservation took root in the same structure of time. According to spiritualists, the past conserved is not anterior to memories recalled. The past persists virtually. As Bergson posited the twin concepts, the actual and the virtual were like obverse sides of the same coin. In an essay that clarified *Matter and Memory*, Bergson elaborated:

> Our actual existence . . . whilst it is unrolling in time, duplicates itself along with a virtual existence, a mirror-image. Every moment of our life presents two aspects, it is actual and virtual, perception on the one side and memory on the other. Each moment of life is split up as and when it is posited. Or rather, it consists in this very splitting, for the present moment, always going forward, fleeting limit between the immediate past which is now no more and the immediate future which is not yet, would be a mere abstraction were it not the moving mirror which continually reflects perception as a memory.[92]

The concept of virtuality was pivotal to explain the coexistence of the *entire* past with the present. Recalling a memory is to make actual a fragment of the virtual past. According to Guyau, "There is nothing like that [a distinct episode] in the brain, not real images, but only virtual images, potentials, that only wait for a sign to pass into an action."[93] Guyau eschewed spatial tropes in order to conceptualize the creative labor of recollection. Memories can be recalled at any moment in the form of an image because the past does not exist else*where*; it constitutes a reserve of experience inhering virtually in the present. Herein lied the constitutive tension that spiritualists articulated between the one and the many—that is, between memory (virtual) and memories (actual).

A conceptual problem came to the fore: How does the virtual become actual? By which means does the past press on the present? For Guyau, there was a creative liaison. In *Genesis of the Idea of Time* (*La genèse de l'idée du temps*)—a book that his adopted stepfather, Alfred Fouillée, published posthumously in 1890—Guyau argued that the recollection of memory (that is, its actualization) involved a qualitative transformation. Elements of the dead past become lively forms of action in the present. The point of view of the actual "is always some scene in space," Guyau wrote, "some event that happened in a material and *extended* milieu."[94] In Paul Ricoeur's succinct reading, "We localize time by localizing in space. This is literally the *mise en scène* of our remembrances."[95] The passage from the virtual to the actual hinged on a dynamism at the heart of the self. As Bergson vividly described it, "We pass, by imperceptible stages, from recollection strung out along the course of time to the movements which indicate their nascent or possible action in space."[96] Recollection thus re-creates the past in the present.

To illustrate the process, Guyau turned to technological metaphors. He likened the brain to a phonograph.[97] The device was a novel invention. Thomas Edison created it in 1877. Other inventors had devised apparatuses for recording sounds, but the phonograph was the first to reproduce recordings. In fact, the phonograph developed out of Edison's trials with telephones.[98] By speaking into the phonograph, vocal vibrations were inscribed in the lines on a metal plate. A small copper disk set the plate in rotation; a stylus ran along the grooves of the lines; and the device amplified the vibrations. According to Guyau, "It is quite probable that, in analogous ways, invisible lines are incessantly carved into the brain cells which provide a channel for nerve streams. If after some time the stream encounters a channel it has already passed through, it will once again proceed along the same path. The cells vibrate in the same way they vibrated the first time."[99] Although neural fibers also repeat the vibrations, there is a noticeable difference. Memories come from the past. "In a similar manner," Guyau wrote, "the phonograph is not capable of reproducing the human voice in all its strength and warmth. The voice of the apparatus will remain shrill and cold; there is something imperfect and abstract about it which sets it apart."[100] Acoustic figures were not new in brain science. Johannes Müller had suggested that "the fibers of all the motor, cerebral and spinal nerves may be imagined as spread out in the medulla oblongata and exposed to the will like the keys of the pianoforte."[101] For Guyau, the phonograph was particularly fitting: memories were like songs virtually conserved on a record. And they were actualized once set in motion. Guyau acknowledged the metaphor's chief limitation. In the phonograph, "there is no transition from movement to consciousness. It is precisely this wondrous transition which keeps occurring in the brain."[102] Similarly, Fouillée recognized that "in Edison's still basic machine, the metal plate remains deaf by itself."[103] Both thinkers saw figuration as a tool to clarify the underlying conceptual problem. It was the task of metaphysics to explain the process by which the virtual passed into the actual.

As for the explanation, Fouillée and Guyau subtly diverged. For Guyau, recalling memories amounted to a knack, "nothing but the art of evoking and organizing these representations."[104] For instance, mnemonic strategies, repetition, and memorization would facilitate recollection. To Fouillée's mind, the key to recollection was practical adaptation. Normally, a barrage of memories vies for our attention. We feel the blank befuddlement of forgetfulness when searching for a name or a verse because we must intervene in the dense struggle of memories—the dynamic texture of selfhood that he likened to "idea-forces." We resist those memories that assail on the mind yet do not support whatever our preoccupation might be.[105] Memories must be adapted to the demands of the moment. As Bergson saw it, "Mr. Fouillée appears to us to have pushed the analysis of Mr. Guyau further." For Bergson, Fouillée identified the seamless "process of *changing*" that memory effectuates between the virtual conservation of the past and its actualization in present recollection.[106] This was the dynamic structure of time that spiritualists developed beyond the spatial categories of brain science. They posited the conceptual integrity of a restless past on the other side of sensory-motor action in the present.

The Telephone and Electrical Configurations of Spirit

The shifting technological climate of France was caught between the past and the future. In the final decades of the nineteenth century, telephones gradually replaced telegraphs. The new device was introduced at the 1878 Universal Exhibition in Paris, where Elisha Grey, Alexander Bell, and Thomas Edison shared the *grand prix* for best invention. Telephones increased the capacity of telegraphic lines by using an undulating current in place of an intermittent or pulsatory current.[107] Numerous signals could be transmitted simultaneously. Furthermore, the speed increased since telephones did not rely on repeaters, wherein telegraph technicians would receive a signal and send it further down the line. Following the First International Exposition of Electricity in Paris in 1881, the installation of telephone

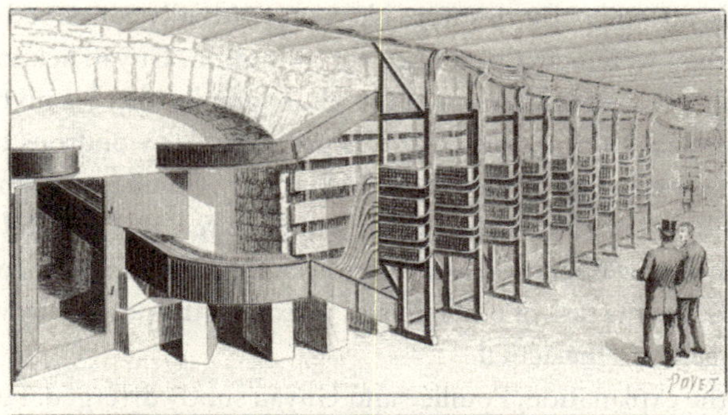

FIGURE 9. Underground telephone cables in Paris, 1895
Source: "Le nouveau système téléphonique de la ville de Paris," *La nature* 23, no. 1 (1895): 40.

wires began in the city. An expansive grid connected Paris's telephone exchanges, which brought private lines to wealthy households (see figure 9).

Like telegraph lines, telephonic cables were installed underground. The network was one of the most advanced and safest in the world. In Paris, cables were not subject to the natural elements. In fact, Parisians were proud of their technological sophistication, particularly when news traveled that exposed lines in New York City came tumbling down in a snowstorm.[108] Moreover, the metallic wires were double. Unlike aerial wires, which made use of a ground current, the subterranean wires comprised an outgoing and a returning line, a configuration adaptable to the nervous system's sensory-motor transmission.[109] The production and installation of telephonic cables fortified a material environment that was literally and figuratively grounded in wires.

An immaterial transformation in the environment took place as well. The experience of time changed radically. Wiring facilitated the synchronization of uniform time, a nearly century-long process during which France, as well as the rest of the world, gradually transitioned from local times (primarily based on the clocks at train stations) to twenty-four-hour standard times zones centered on the prime meridian.[110] Telegraphy had led the way. The completion of the second transatlantic telegraph cable in 1866, a more durable replacement of the first attempt from a decade prior, made it possible not only to send electric signals between Newfoundland and Ireland, but also to clock those signals' distance and speed. The cable set in motion the measurement of the earth's longitude—each time zone spans, in theory, fifteen degrees longitude. The drive toward temporal unification created an experiential rift. The natural experience of local time, premised as it was on the rhythm of the day, gave way to the abstract conventions of time implemented by administrative authorities. Before, time differences between cities had been as little as 15 minutes apart. But a national mean time—"Paris time"—was put in place on French railways in 1889. It was only in 1911 that France finally adopted the Greenwich meridian as the basis for a uniform time.

Telephony accelerated the pace of life. The impression of distance contracted as the experience of simultaneity expanded.[111] The speed of telephone conversation felt instantaneous in contrast to the telegraph. When it came to metaphors of the brain, however, the technological change bore little fruit. Nervous "wiring" continued to evoke the image of telegraphic circuits, even when the telephone was deployed as the rhetorical figure of choice. The memory of telegraphic circuitry lingered in the cultural imagination. In fact, the telephone was first known as a "harmonic telegraph," the name Alexander Graham Bell gave to his 1876 patent. Telephone lines depended on the same infrastructure of telegraphs. In the human sciences as well, telephonic metaphors carried the trace of their telegraphic predecessor.

Telephonic metaphors served to explain the conservation and recollection of memory. Yet the figure was often anachronistic. The implied technology was more fitting of a telegraph, which sent electrical pulsations (and not an undulating current across multiple ranges) between stations. The French neurologist Maurice de Fleury (1860–1931) invoked the telephone to explain his associationist model of vocal memory:

> A syllable is constantly repeated to the child: each time, that sonorous wave sets vibrating the terminations of the auditory nerve in the ear, which it follows to the point where it ends in the cerebral cortex (A). But that vibration is always tending to escape. It is a force that has come into us and wants to go out here as elsewhere, sensation demands to be changed into action, and the first part of the reflex, centripetal vibration, commands the second, centrifugal vibration. The nerve wave does not

> then stop in (A), except to leave the memory of it; it goes on its way and, following the fibres of association (A–M), like the wire of a telephone, it ends in (M).[112]

The accompanying diagram accentuated the communications trope (see figure 10). Fleury drew straight lines—a representation suggestive of a cabled network—in order to depict the connections among memory centers. For Fleury, wiring also entwined society and nature. His overarching vision was of a world in which "man seems to be bathed in an ocean of various vibrations which are changed, on contact with his nerves, into nervous vibration, and, under that form, reach the grey matter, the place of consciousness."[113]

Bergson deployed a similar figure in *Matter and Memory* to opposite ends. He likened the brain to a "central telephone exchange," which was responsible for directing calls in a limited geographic area. Workers sat at a switchboard and connected subscribers' lines. Exchanges made it possible to call without direct lines between parties. For Bergson, the exchange signified the hub of the body's sensory-motor circuitry. "The truth is that my nervous system, interposed between the objects stimulating my body and those that I can influence, plays the simple role of a conductor that transmits, allocates, or inhibits movement."[114] Bergson took the figure to mean that the brain did not house memories, which could therefore not be localized in neural tissues. The organ facilitates only recollection in the present and not the conservation of memory in the past. In the words of Gilles Deleuze, interpreting the Bergsonian metaphor, "The brain was only an interval [*écart*], a void, nothing but a void, between a stimulation and a response."[115] For Bergson, it followed that lesions to the brain inhibited solely the actualization of memories. Cerebral damage had no effect on past, virtual memories.

Perhaps Bergson borrowed the figure from Hippolyte Taine. For the positivist thinker, wiring metaphors emblematized the marriage of technology and neurology. In his 1870 opus *On Intelligence* (*De l'intelligence*), Taine argued that scientific progress had brought about a post-metaphysical world without need for immaterial notions of spirit. In his personal copy of the book, Bergson scrawled few notes. But he did mark two lines where Taine figured the brain as a "telegraph exchange," whose operations are like a "*repeater and multiplier*, in which the diverse departments of the gray matter fill in the same functions."[116] The terms were fashioned for wired communications: repeaters linked telegraph networks to a hub; multipliers increased the voltage of an electrical current. Taine invoked the figures to bolster his reductionist ontology. He believed that ideas were not created

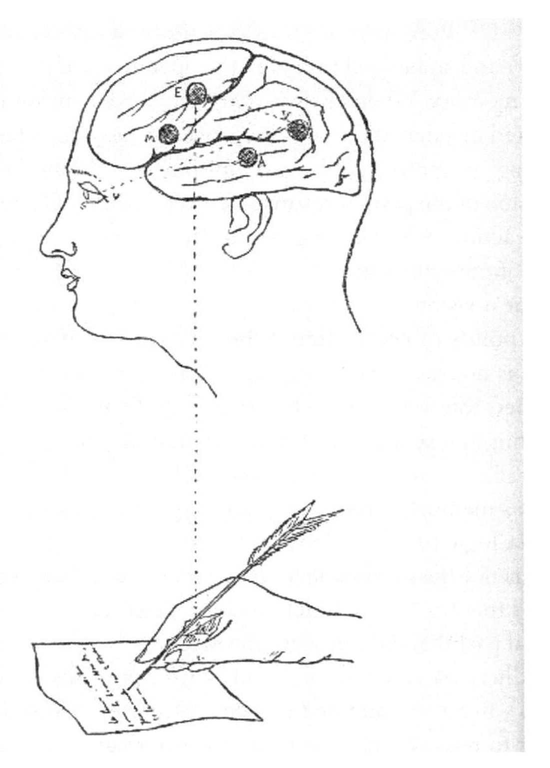

FIGURE 10. Fleury's memory diagram in *Introduction à la médecine de l'esprit*
Source: *Introduction à la médecine de l'esprit* (Paris: Félix Alcan, 1898), 162.

ex nihilo but were epiphenomena of the nervous system's electrical signals. That the spiritualist Bergson shared a technological idiom with such an unapologetically anti-metaphysical thinker testifies to their proximity in an intellectual community that was sustained in no small part by the circulation of technological tropes.[117]

Bergson also meant for the telephone metaphor in *Matter and Memory* to illustrate the interrelation between two poles of memory: action-oriented perception in space and what he called "pure memory" in time. On the one hand, dreams offered a glimpse into pure memory. He turned to the French writer Alfred Maury (1817–92), who argued in *Sleep and Dreams* (*Le sommeil et les rêves*) (1861) that memories occupied the entirety of the sleeping mind.[118] From descriptions of dream states, Maury suggested that dreams take on a life of their own, freed as they are from the sensory-motor prerogatives of wakefulness. Bergson drew from the book the critical insight that pure memory is not anarchic. Its structure retains the entirety of the past.[119] Pure memory thus constitutes a reserve of the virtual. On the other hand, aphasia illuminated the operations of perception severed from pure memory. Whereas dreaming unlocked a surfeit of memory, aphasia resulted in a deficit of memory. Since the disorder affected only the neural "wiring" of motor-memory, symptoms were limited to the superficial dimension of the past. Bergson took this to mean that aphasiacs—untethered from action-oriented language in the present—descended into an inner state of pure memory.

The division between perception and pure memory served to elucidate their points of connection. When it comes to perception in the present, there is no *selection* of image-memories from the depths of pure memory. For Bergson, we open up to the arrival of particular memories by actively resisting the sensory-motor orientation of perception. Since the body is naturally oriented toward acting on the material world, perception severs useless memories from our practical preoccupations. Against the associationist logic of psychology and neurology, Bergson argued: "*Association*, then is not the primary fact: *dissociation* is what we begin with."[120] He reasoned that the brain's functions were negative. It *suppressed* aspects of the virtual past that did not serve the present.[121]

Adhering to the same method as in his previous book, *Time and Free Will* (1889), Bergson examined two dimensions of memory at their extremes in order to reconstruct their nexus in quotidian experience. We employ perceptive skills while going about mundane tasks, touching only the shallows of memory. In exceptional moments, we undergo an intellectual endeavor, think hard, and maybe even philosophize. In so doing, the mind plumbs the bowels of pure memory. Perhaps we struggle to recall the name of a friend seen long ago, or pose a difficult math problem for which rote arithmetic will not suffice. In such instances, the two dimensions of memory come together in volition. The degree to which memories mix with perception in the present depends on how willfully we arrest our fixation on the moment at hand and plunge into the past. In this virtual reserve of time, all prior experience hangs together in what Bergson called the "planes of memory." Each plane constitutes an effortful dilation of the *whole* of one's past. An exertion of thought opens ever deeper on each plane. Bergson described these planes as if they were "a nebulous mass, [which] seen through more and more powerful telescopes resolves itself into an ever greater number of stars."[122] He noted that the each plane is not quantitatively greater than another; they "are certainly not easy to define, but the painter of mental scenery may not with impunity confound them."[123] The spiritual strata of time cohere qualitatively.

Why, then, do certain images emerge from pure memory rather than others? How do we come to recall any memory at all? According to Bergson, opening our mind onto the outpouring of the past involves a mental act of attention. He again insisted that attention depends on an exertive resistance to the body's orientation toward action. "Always *inhibited* by the practical and useful consciousness of the present moment, that is to say, by the sensori-motor equilibrium of a nervous system connecting perception with action, this memory merely awaits the occurrence of a rift between the actual impression and its corresponding movement to slip in its images."[124] In the gaps between stimulus and response, the nervous system frees room for memories' intervention. For Bergson, there are two ways that image-memories intervene: "Essentially fugitive, they become materialized only by chance, either when an accidentally precise determination of our bodily attitude attracts them, or when the very indetermination of that attitude leaves a clear field to the caprices of their manifestation."[125] Both avenues follow the contingent connections between actual matter and virtual time.

Herein lay the conceptual payoff of the telephone metaphor. To actualize virtual reserves of the past depends on a suspension of the brain's telephonic disposition. Technological figures were doubly useful to illustrate the process. Bergson not only deployed the central telephone exchange to clarify the functions of the brain; he also drew on electrical systems in order to explain the process by which the images that bubble up from pure memory become grafted onto motor activity in the present.

Bergson depicted attention as a series of billowing circles (see figure 11). The inner circle (A-O) is present perception. It corresponds to the neurological activity by which the object perceived (O) leaves an immediate impression on the body's sensory organs (such as when looking at a sunrise). The outer circles represent the progressive dilation, as it were, of perception's aperture. Together, they represent "a closed circle in which the perception-image, going toward the mind, and the memory-image, launched into space, career the one behind the other."[126] Bergson further characterized B, C, and D as "efforts at intellectual expansion."[127] Each enacts a heightened resistance to the practical demands of the outside world. B′, C′, and D′ are deeper image-memories that return to imbue the object perceived with elements of the past. He repeated the claim that this does not involve a selection of discrete images. "If perception [A-O] evokes in turn different memories," Bergson wrote, "it is not by a mechanical adjunction of more and more numerous elements which, while remaining unmoved, it attracts around it, but rather by an expansion of the entire consciousness [B, C, D], which, spreading out over a larger area, discovers the fuller detail of its wealth [B′, C′, D′]."[128] It is worth emphasizing that this seamless act of attention takes place within the object perceived. He in-

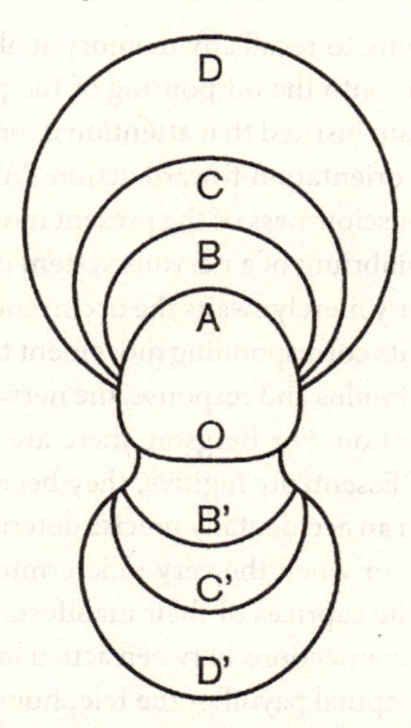

FIGURE 11. Bergson's diagram of attention in *Matière et mémoire*
Source: *Matière et mémoire* (Paris: Félix Alcan, 1896).

sisted, "the formation of memory is never posterior to the formation of perception; it is contemporaneous with it. Step by step, as perception is created, the memory of it is projected beside it, as the shadow falls beside the body."[129] One does not depart from the present and go elsewhere. The virtual surges into the actual.

It is significant that Bergson described attention as a *circuit*: "reflective perception is a *circuit*, in which all the elements including the perceived object itself, hold each other in a state of mutual tension as in an electric circuit, so that no disturbance starting from the object can stop on its way and remain in the depths of the mind: it must always finds its way back to the object whence it proceeds."[130] If we compare Bergson's diagram to that of a basic electrical circuit, the resemblances—both conceptual and visual—are quite noticeable. An electric circuit flows between two unequally charged surfaces, one with a positive valence and the other with a negative valence. Electrons pass from the positive surface and return via the negative surface, as seen in the diagram from a late nineteenth-century electronics manual (see figure 12).[131] Typically, an electromagnet or battery would generate the same current traversing the circuit. The effective force leaving the battery depends on two things: (1) the *tension* of the battery (more often referred to as *voltage*), which sets the current in circulation; and (2) the *resistance* that the current encounters while traveling across the circuit (measured in

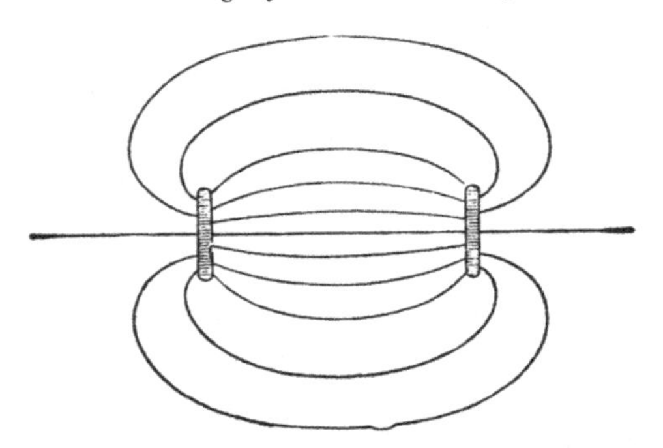

FIGURE 12. Diagrams of electrical fields
Source: T. O'Conor Sloane, *The Standard Electrical Dictionary* (New York: Norman W. Hensley, 1892), 234.

ohms). Bergson drew on both concepts. Attention *resists* the demands of the present moment in order to expand the whole of memory, whose different planes correspond to "the degree of *tension* which our mind adopts."[132] Further, when a charge moves across a distance, it generates an electrical field. Normally, when there is no physical contact between charged surfaces, electric force travels in straight lines between each surface. But this is the case only if the surfaces are close enough. When space comes between charged surfaces, the lines of force arch outward and generate an electrostatic field, as seen in the diagram.[133] In this case, electricity spills onto the environs. Notice how this representation of the phenomenon in an electrical manual from the period bears a noticeable similarity with Bergson's diagram for attention.

Bergson did not cite an electrical manual. Nonetheless, I believe it is reasonable to conjecture that electrical circuitry provided a figurative template for Bergson to conceptualize the structure of attention. The lines function analogously to Bergson's billowing circles, which extend outward from the perceiver and return back to the object of perception. The electrical figure illustrated that the exercise of attention always follows the same path. With greater distance between stimulus and response (like the wider separation between charged surfaces), attention strays from the proximate lines of physiological perception and extends along the "atmospheric" lines of pure memory—what could be likened to the mind's electrical field.

This metaphoric context enriches our understanding of Bergson's claim that pure memory is not confined to the space between our ears. Perhaps we could say that the virtual is distributed across the cultural and technological climate. The wired environment—in which the French social sphere took shape—was an appendage of the self. In imagining the conservation of the past to inhere across a milieu that enveloped the brain, Bergson set himself apart from his spiritualist contemporaries. He advanced the most capacious understanding of spirit, an understanding that set the play of time and space in touch with the inner and outer bounds of selfhood.

Bergson acknowledged that *Matter and Memory* was "frankly dualistic."[134] His primary interest was the mind-body problem. In the conclusion of the book, he argued that the connections between embodied action in the material world and pure memory in the depths of spirit would remain an intractable riddle so long as they were conceptualized spatially, "like two railway lines which cut each other at a right angle."[135] By instead conceptualizing the dimensions as temporal processes, "the rails come together in a curve, so that we pass insensibly from one to the other."[136] Technological metaphors were crucial tools that Bergson mobilized in order to reframe static oppositions between spirit and matter as a dynamic transformation by which spirit becomes materialized.

Bergson went on to expand his figurative tool kit. He construed the brain as an "athletic organ" (*organe sportif*), a figure that emphasized the organ's propensity toward sensory-motor action.[137] Figuration also allowed Bergson to elucidate the cerebral points of contact between matter and spirit. In *Creative Evolution* (*L'évolution créatrice*) (1907), Bergson suggested, "The consciousness of a living being . . . is inseparable from its brain in the sense in which a sharp knife is inseparable from its edge: the brain is the sharp edge by which consciousness cuts into the compact tissue of events, but the brain is no more coextensive with consciousness than the edge is with the knife."[138] The cerebral and the spiritual were figured in tandem. But their respective realms were not. In *The Creative Mind* (*La pensée et le mouvant*) (1934) Bergson clarified, "the brain does not have thinking as its function but that of hindering the thought from becoming lost in dream; it was the organ of *attention to life*."[139] Elsewhere, he referred to "the brain, organ of pantomime," whose operations serve to "internally activate the life of spirit."[140]

Although his engagements with cerebral localization and psychopathology were peculiar to the state of each field at the turn of the century, Bergson's insight was prescient. Figures have been indispensable in the neurosciences. The central telephone exchange gave way to an information-processing system, a cybernetic machine, and eventually a cognitive computer. Today, it's commonplace to think of the brain as a neurochemical "plastic" organ. As much as scientific precision eschews metaphors, they have not been so easy to eliminate; indeed, figuration has remained inescapable—and deeply informative—in the history of the brain sciences.

❋

In fin de siècle France, figures of the brain made the organ the locus as well as the instrument of the self. Cultural imagery suffused laboratory science and clinical medicine as the problems of brain research circulated throughout society. Technological rhetoric provided a shared frame in which neurologists, psychologists, and philosophers conveyed and contested the problems that arose from the inescapable intersection of the spatial and temporal dimensions of memory. Herein lay the thrust of spiritualist thinkers' conviction that memory—integral as it was to the flux of time—was not localizable. Memory could not be found in the brain for the same reason that time did not unfold in the space behind our eyes. For Bergson, Egger, Fouillée, and Guyau, metaphors of inscription, communications, and industry showed that temporality was far more vast—indeed, far more pliable—than the finite human body. These thinkers showed, moreover, that cerebral localization and psychopathology did not advance through linear progress, but carried forth their own restless memory. Far from abandoning immaterial notions of selfhood to the past, the young brain sciences continuously regenerated the problematic relations of matter and spirit.

✷ CHAPTER 5 ✷

The Institutions of the Intellect, or Spirit contra Kant

"All those following the contemporary intellectual movement with some attention agree that philosophical speculation has been undergoing for some years now, at least in France, a new orientation."[1] Thus opened a scathing 1898 screed written by Baptiste Jacob (1858–1909), a philosophy instructor at a lycée in Saint-Brieuc, a small town in the northwest corner of France. In the final years of a long career spent teaching the state philosophy course, Jacob found himself baffled by the ideas of selfhood and science that had gained popularity in his profession. It was not that he failed to grasp advancements in neurology and psychology. Jacob saw himself among Immanuel Kant's inheritors for whom scientific inquiry depended not only on the exploration and collection of facts in the material world, but also on the conceptual contributions of a thinking subject. By reflecting on the labor of concepts—the rational activity by which we make the world meaningful—philosophy would clarify, refine, and ultimately fortify the sciences' foundations. "Above scientific laws arise the first principles which are the object of philosophy."[2] But instead, a strange, "new orientation" had taken hold: what Jacob called "neo-materialism."[3] Not content to reflect on the concepts organizing scientific inquiry, this new orientation instead took the sciences to depend on the pragmatic interests of an embodied subject navigating the material environment.

The charge of "neo-materialism" was revelatory. The accused never ascribed the label to themselves; yet, it was hardly a misnomer. Jacob distilled the thrust of the materialist current inflecting spiritualist thought. Unlike reductive materialisms, which conceptualized matter as the substratum and final explanation of spirit, this "neo-materialism"—and its leading practitioner, Henri Bergson—reimagined matter to enter into a partnership with the spiritual powers of memory, creativity, and action. What's more, Bergson and his peers made the case that intellection alone, unmoored from our fleshly life, was a blunt instrument. The view was anathema to Jacob.

Although he also rebuked the reductionist and determinist claims of crude materialists, Jacob believed that the self's spiritual powers were accountable to reason. He invoked the tradition of German idealism and translated its central concepts into his native language. For Jacob, the locus of *l'esprit* was the understanding (*Verstand*)—not corporeal activity. Motivating his account of spirit was a vision of philosophy's epistemological purpose that had developed out of the comingling of French and German thought across the nineteenth century: to be the handmaiden of the sciences; to shore up their claims; to ensure, as Jacob put it, "their intelligible foundation and their ideal model."[4] Jacob believed that philosophy was in the service of science. For Bergson, it was science that served philosophy.

Contestations over science, spirit, and intellection came to the fore at the turn of the twentieth century. Spiritualist thinkers encountered a constellation of critics who ushered Kant's teachings into a new era. These neo-Kantians rekindled an epistemology originally elaborated in the late eighteenth century (when geometry and physics were taken to be paradigmatic sciences) and adapted it to the scientific and social exigencies facing modern France.[5] At stake in these contestations were two entwined problems. The first concerned competing visions of spirit; the second was philosophy's relation to the sciences. My aim in this chapter is to explore these braided disputes and to explain how they fueled a reworking of spiritualist thought: namely, the "neo-materialism" that took hold across the institutional landscape of France in the decades of the 1890s through the 1900s. Ultimately, the most celebrated achievement of the period was Bergson's opus *Creative Evolution*, in which the author exposed the limitations of intellection from the vantage of evolutionary biology.

Although originally expounded in Germany, Kantian thought came to flourish in France by the end of the nineteenth century.[6] It found ardent and widespread support as a reaction to the growing materialist inflection in ideas of spirit. Bergson and others questioned the relevance of Kantian thought. In their eyes, the working categories of the mathematical and physical sciences had become sclerotic, no longer capable of resolving the problems facing the human and life sciences. But the spiritualist critique also cut deeper. It upended the foundations of Kantian epistemology: in particular, the architectonic division between the realms of nature and freedom. The *Critiques* of pure and practical reason had segregated the laws determining humans' representations of the natural world from the laws regulating humans' moral conduct in the social world. Causality explained the recurrent relations found in matter; norms guided conduct among ethical beings. For Kant and his followers, the division served to safeguard autonomy. Humans could not locate the source of their autonomy in nature because freedom did not constitute an empirical object—the kind of thing uncovered by observation and experimentation. Humans exercise their freedom, Kant held, by transcending the natural world. We employ a priori concepts to make judgments (*Urteilskraft*)—the definitive act of intellection, by which particular facts are subsumed under general concepts. Despite their shared commitment to defending autonomy, spiritualist thinkers chipped away at the scaffolding on which Kant erected humans' freedom. They believed that neurophysiology had shaken the pillars separating nature from freedom and, as a result, the Kantian primacy of the intellect tumbled as well. For the thinkers in this chapter, autonomy was best defended by exploring its efficacy in the natural world; they meticulously described the experiential contours of the self's coping with materiality. Science, it followed, was more than applying concepts. It reflected humans' efforts to make use of matter; to mold it to their purposes; to satisfy the body's pliable demands.

This account of selfhood and science left Jacob perplexed when he read Bergson's 1896 essay in the *Philosophical Review of France and Abroad* (*Revue philosophique de la France et de l'étranger*). The journal was the chief organ of philosophy in France. Articles written by academic philosophers appeared alongside others by physiologists, biologists, and neurologists. In a piece that became part of *Matter and Memory*, Bergson offered a review of pathological studies in neurology and psychology.[7] As I showed in the previous chapter, Bergson delineated the corporeal points of contact by which memory entered into and became actionable through the brain. This liaison between spirit and matter left Jacob with more confusion than clarity. Most of all, the pillars segregating nature and freedom were left lying in rubble. Jacob scribbled his reaction in a letter to a friend, "At the moment when all the psychology on which I had grounded my interpretation of things is undermined at the base, I would like to have you for an ally."[8] When it came time to publish his critique, Jacob sought a forum other than the *Philosophical Review*. He desired a community of like-minded readers who were intent to shield the spiritual from the material.

So Jacob approached a new journal. The *Review of Metaphysics and Morals* (*Revue de métaphysique et de morale*) was launched in 1893.[9] Its appearance marked a momentous event in French culture. In contrast to the wide interdisciplinary range of the *Philosophical Review*, the new journal was intended to be an exclusive forum for academic philosophy. The founders were three young men, Léon Brunschvicg (1869–1944), Élie Halévy (1870–1937), and Xavier Léon (1868–1935), who shared a reverence for the progress of science and the authority of reason. They found Jacob's submission to level a much-needed blow to the spiritualist cause. He forcefully charged, "The philosophy of Mr. Bergson is . . . of its time; it comes at a moment in a social dialectic, and one could, it seems, define it without injustice as the metaphysical expression of two principal forms of our 'anxiety': mysticism and impressionism."[10] The accusation of "mysticism" impugned the metaphysics of spirit, whose powers were thought to be energetic, effortful, and not primarily rational. "Impressionism" alluded to Bergson's claim that immediate experience constitutes a continuous flux.[11] As an antidote to the charges, Brunschvicg, Halévy, and Léon convened a journal wherein readers and writers found refuge from the "irrationalism" of "neo-materialism." The neo-Kantian orientation of the *Review of Metaphysics and Morals* buttressed the epistemological segregation of nature and freedom. For the founders, rational thought served to refine scientists' working models, to parse the legitimate from the specious, and eventually to bring knowledge about the material world into harmony with the totality of human experience.

The founders did not wish for philosophy to be hermetically sealed from French society. Brunschvicg, Halévy, and Léon shared the ambition to create a venue of conversation in which a philosophical elite would buoy the democratic values of the Third Republic. They sought to put the state in service of the moral development of citizens.[12] Although their work did not espouse an explicit political ideology, it was not laissez-faire liberalism.[13] The viability of Republican culture hinged—so the founders believed—on reflective, upright citizens whose exercise of reason expressed not the self-interest of market actors, but a historically rich appreciation of science, morals, and politics. They saw the state philosophy course as the best avenue to bring these domains into harmony. Halévy, Léon, and Brunschvicg articulated in the journal's manifesto that its goal was to reinforce the *light of reason* (*lumière de la raison*), implying that spiritualist thought had left French culture in darkness. Articles illuminated the universal principles motivating the country's scientific and social progress.[14] The *Review of Metaphysics and Morals* would thus shepherd Kant's project into modern France.

As it turned out, the *Review* went on to publish the second of Bergson's articles to become a part of *Matter and Memory*.[15] Why would the journal feature the work of an intellectual foe? Being the aspiring university students that they were, the founders ceded to the practical exigency that their journal's success relied on support from the French academic establishment. Moreover, their aims were not dogmatic. Despite the founders' neo-Kantian persuasion, the journal provided a forum open to diverse views. It was in this milieu, I want to suggest, that spiritualist thought was elaborated, sharpened, and radicalized thanks to productive conceptual tensions with neo-Kantianism. The result, according to Bergson, was a higher stage of spiritualism: "positive metaphysics," as he called it.[16] In the words of Édouard Le Roy, it was "spiritualist positivism," which, "far from having been called from outside, as it were, by metaphysical and moral preoccupations . . . has appeared from the inside of science, under the pressure of its internal needs, and in contact with its very facts and theories."[17] Alfred Fouillée characterized the moment as "positive and experimental."[18] This cluster of names converged in a community of readers and writers where a sustained debate unfolded over the meaning and role of the intellect. Across the two decades that straddled the turn of the twentieth century, spiritualist thought followed a path, in the words of a prescient critic, leading to "neo-materialism."

The Review of Metaphysics and Morals *and the Workshop of French Neo-Kantianism*

When the *Review of Metaphysics and Morals* burst onto the scene in 1893, it was France's first genuinely academic philosophy journal in a double sense. First, articles were written by professional philosophers who held teaching positions as civil servants. Second, the journal's founders came of age in the new academic institutions of the Third Republic. Brunschvicg, Halévy, and Léon were among the earliest generation to be inculcated by an official Republican curriculum. They attended the prestigious Lycée Condorcet in Paris, where the mandatory philosophy course was taught by Alphonse Darlu (1849–1921). Although he published little, Darlu had a profound influence on his students. They credited him with having imparted the reverence for rationality, the "divine spark," behind the journal.[19] The founders were graduated in 1888, after which they went on to study philosophy up the road—Brunschvicg and Halévy at the École normale supérieure; Léon, a stone's throw away at the Sorbonne.[20] Together, they launched the *Review*, no small feat for young students ensconced in the delectations of university life.

The two operative words in the title, *metaphysics* and *morals*, lent expression to the founders' philosophical orientation. Yet, "metaphysics" seemed to flirt with the spiritualism whose "irrationalism" they so vehemently decried. Léon insisted that the word be included. He wanted to distinguish the *Review* from positivists' skepticism for all forms of knowledge not rooted in empirical data. The founders conceived of metaphysics as the inquiry into the basic principles organizing thought. Halévy wrote in a letter to Léon that it was acceptable in this strict sense: "Having reflected on it all . . . I withdraw my objections to the word: metaphysics, if you mean by that not a separate science, discussed in isolation of a determined number of special problems, but a method, opposed to the *positivist* method of observation."[21] As for "morals," the word captured the founders' aspiration to offer a model for French social life. To their mind, Kant's three *Critiques* provided the blueprints for rational ideals of citizenship under the Third Republic. Léon laid out his compatriots' vision in a private letter: "If we were permitted to summarize in two formulas the spirit of our direction, we would willingly say that the *Review of Metaphysics and Morals* pursues a double goal: that of bringing together the philosophy of the sciences on terrain other than that of facts, that is on the terrain of ideas; and that of bringing together the philosophy of morals on terrain other than that of faith in sentiment, that is on the terrain of reason."[22]

Each of the founders went on to achieve success in his lifetime. Léon spearheaded the journal and served as editor until his death in 1935. Along the way, he published numerous philosophical essays, including *The Philosophy of Fichte* (*La philosophie de Fichte*) (1902), which he expanded into a three-volume series during the 1920s.[23] Halévy came to be well known for his magisterial series on British utilitarianism and a history of nineteenth-century England. He achieved public prominence in Paris, giving biannual lectures on the history of socialism at the École libre des sciences politiques between 1902 and 1937.[24] Among the group, Brunschvicg enjoyed the most prolific philosophical career. After teaching the mandatory philosophy course at provincial lycées, he took a professorship at the Sorbonne in 1909. There, he published a wide-ranging corpus in defense of neo-Kantian thought. For Brunschvicg, the intellectual basis of selfhood lay in the act of giving meaning to the world. He staked out his position in an early article, "Spiritualism and Common Sense," in which he reclaimed the idea of spirit from spiritualism: "To think is to have a spirit, from the moment that it becomes conscious of itself, conceives itself as an indefinite and single activity."[25] For Brunschvicg, the nature of spirit was rooted not in the finite conditions of embodiment (*pace* Bergson and fellow spiritualist thinkers) but in the infinite power to impart significance to the objects of thought.

Brunschvicg, Halévy, and Léon were Jewish. Although they were assimilated into French culture, and did not strictly adhere to Halacha, the founders were well aware that they stood out in Catholic France. Such was especially the case during the late 1890s, when the Dreyfus affair stirred anti-Semitic fervor. The three thinkers supported the cause of Captain Alfred Dreyfus, the Jewish artillery officer who was falsely charged with treason in 1894 and sentenced to life imprisonment. In 1906 he was exonerated. Nonetheless, Brunschvicg, Halévy, and Léon encountered the vitriol of French conservatives. One historian suggests that the affair impacted the founders so forcefully that the journal came to serve as their outlet to escape the tensions between nationalism and Zionism that characterized the politics of the period.[26] Neo-Kantian philosophy offered a cosmopolitan avenue out of the dichotomy. In the founders' eyes, reason was not the purview of any cultural particularity. They believed the solutions to the country's political crises hinged on the cultivation of a critical attitude premised on rational thought.

Prior to the *Review of Metaphysics and Morals,* two French journals had been dedicated to philosophy: the *Philosophical Review* and *Philosophical Critique* (*Critique philosophique*). The former was more widely read. It was founded in 1876, a creation of the young Théodule Ribot, and it rapidly won the support of the academic establishment. The *Philosophical Review* cemented itself as the primary arena for philosophy alongside the human sciences. In fact, Ribot was so convinced of his journal's prestige that in 1893 he questioned the viability of the inchoate *Review of Metaphysics and Morals.* "Have you received the first issue?" he asked in a letter to a colleague. "It will be a closed *Review,* strictly academic, *Sorbonnienne,* and *Ravaissonnienne.* I doubt that it will recruit many subscribers and that it will damage us."[27] Later, Ribot elaborated on his skepticism, describing the founders as "young people, rich, Jewish, and very metaphysical."[28] Such were the words of a scientific thinker by then well respected in the academic establishment.

Philosophical Critique represented the margins of academic life. Charles Renouvier (1815–1903) had created the journal along with François Pillon (1830–1920) in 1867. Both were independent philosophers without university posts. *Philosophical Year* (*Année philosophique*), as it was originally titled, was a forum for the two editors to repackage Kantian thought for French audiences. Their aim was to delineate the bounds of faith and reason in the face of growing secular forces across Europe. During the Franco-Prussian War, publication was suspended. Soon after, the journal relaunched with the apt *Critique* in the title. The editors wrote, "Germany, the homeland of Kant, but which has tasted Kant only as a repulsive scholastic, seems even more distant than France now to enter into the mother thought [*pensée mère*] of criticism and to embrace its moral assertions."[29] Critical philosophy would overcome the grand debates between rationalism and empiricism, as Kant had originally presented his system. Renouvier's ambition was also that criticism would open a public space of debate in which readers would relinquish dogmatic faith and advance the rational reorganization of society originally set in motion by the French Revolution.[30]

Renouvier had initially formulated what he called "new criticism" in a voluminous body of work, *Essays of General Critique* (*Essais de critique générale*) (1854–64). Renouvier deeply admired Kant, but his reading of the three *Critiques* was idiosyncratic. His *Essays* shared the phenomenalist commitment that reality is meaningful only in relation to human thought. Reason governs the logical relations among the objects of experience but offers no guarantees about their existence in the material world. With Kant, Renouvier agreed that reason could not be found in nature. Against Kant, Renouvier did not see reason as an impersonal authority. In the first *Critique,* Kant had distinguished two senses of the self: empirical self-consciousness (comprising the sensual feelings of one's own thoughts, emotions, and desires) and transcendental apperception, the logical form by which one determines the relations of experience. Kant associated reason with the latter. As he argued in the Transcendental Deduction, "There can be in us no modes of knowledge, no connection or unity of one mode of knowledge with another, without that unity of consciousness which precedes all data of intuitions, and by relation to which representations of objects is alone possible."[31] This rational sense of the self was the result of deductive reasoning. Kant took the intuitive fact that all experience is one's own and derived what must be the case for it to be possible: that experience presupposes a logical form; the "I think" accompanies all our representations. Renouvier rejected the transcendental sense of selfhood for being an abstract presupposition of any experience whatsoever. In so doing, he introduced an enduring tenet of French neo-Kantianism: that selfhood is inseparable from a situated subject living in a concrete context; that the conditions of experience are dynamic, forged through the warp and woof of history.

Renouvier offered an ethical interpretation of Kantian idealism that placed the will at the helm of reason. Two consequences followed. First, Renouvier held that the process of concept formation depends on individual belief. Experience admits unity only if one believes in the self's capacity to reason. And the ground of belief lies in the will. His understanding of the will had long-standing roots in the French language. *Libre arbitre* characterizes freedom as a choice. Free choice depends on the reflective capacity to represent possible options, weigh their potential consequences, and pursue a course of action. This legalistic sense of the will figures the self as a judge (following the Latin, *arbitrium*) who renders a judgment, as it were, in the court of spirit.[32] As he characterized it, "we collect our forces, we fix our sentiments, and it is only then that—marching straight along the path of life we have chosen, we become our true selves. To thus transform plurality into unity is to make contact with ourselves in order to place the government of our innermost spirit in the hands of a single idea."[33] In Renouvier's eyes, the proof of freedom was that one *chooses* to believe it.[34] Despite the prominent role that he afforded the will, what distinguished his thought from that of spiritualists was that Renouvier did not take willing to reflect *la volonté*—that is, an embodied and energetic activity, which involves not simply the rational power of thought (*cogito*), but more importantly, an inclination or desire (*volō*). This tenet of French neo-Kantianism ensued: that the will is motivated primarily by reasoning and not by exertion.

The second consequence was that Renouvier parted with Kant's commitment that the will inheres in the noumenal self. For Renouvier, willing finds expression in reflection, deliberation, and judgment. He argued that by preserving the integrity of free will outside experience, Kant had rendered the animating power of our moral life ineffective in the social realm. As a critic summarized, "When we speak of human liberty, we speak of the liberty of a man who is known to us, a man who is placed in time and space and not in the spontaneity of a being who escapes all knowledge."[35] Renouvier elevated Kant's second *Critique* above the first. The exercise of practical reason made possible the legislations of pure reason. The brightest of Renouvier's disciples, Octave Hamelin (1856–1907), expounded on the psychological role of willing in reason. For Hamelin, the will "is produced in all these higher states of consciousness that are called attention, systematic abstraction, sustained and varied reflection, states which are genuine 'self-motivated analyses.'"[36] Hamelin was part of a team of writers whose work for the *Philosophical Critique* espoused optimism in personal autonomy and democratic authority. According to their "new criticism," the individual will was the engine driving the collective will of the French nation-state.

The founders of the *Review of Metaphysics and Morals* stood on the shoulders of their neo-Kantian predecessors. Renouvier, Pillon, and Hamelin had popularized a conception of spirit predicated on the individual self. In turn, Brunschvicg, Léon, and Halévy understood philosophy to be a rational enterprise that refined society's systems of meaning. For these thinkers, reason served as the sturdy basis of morals as well as the sciences, which—they argued—were open to the shifting and conditional character of the world. The aim of scientific inquiry was not to uncover a structure underlying nature, but to pose, test, and revise hypotheses. A frequent writer for the journal, Lionel Dauriac (1847–1923), noted in a letter to Renouvier, "I would never have believed it long ago; now I'm persuaded that if the *Critique* had never come about, then the *Review of Metaphysics and Morals* would not have been born."[37] Dauriac hastened to add, "The philosophical extravagance has admittedly been lost."[38] The journal's founders sought to salvage Renouvier's phenomenalism but also to purge his emphasis on the psychology of belief. Their strain of neo-Kantianism was stridently rational.

In fact, the opinion that Kant's reception had not been sufficiently rational was shared by many readers of *Review of Metaphysics and Morals*. Reflecting on the fate of neo-criticism, Hamelin wrote to Renouvier, "Certainly, I do not purport to be a perfectly orthodox disciple because I strive to do greater than you had ever wanted on behalf of reason in consciousness."[39] To identify spirit not with individual belief, but with the social advancement of reason: this was the new journal's goal. Léon put it succinctly. He characterized the *Review of Metaphysics and Morals* as an alternative to the preexisting philosophy journals of the nineteenth century: *Philosophical Review* and *Philosophical Critique*—that is, as an "anti-Ribot and Renouvierian journal."[40]

When the Universal Exposition returned to Paris in 1900, the editors of the *Review of Metaphysics and Morals* made the most of the occasion. Building on the journal's success, they organized the first International Congress of Philosophy. The event drew hundreds of philosophers from Europe and America. It could be said to have brought the organizers' neo-Kantian vision to fruition: a multinational public gathered to deliberate over the organizing precepts as well as the future orientation of social and scientific thought. Léon served as the secretary along with Halévy, whose position was treasurer. Darlu took over the role of vice president.[41] It was Émile Boutroux who presided. Representing French academe, he was fitting not only to inaugurate the congress but also to convey the prestige of the nation's philosophy (while cultivating courteous relations with peers beyond the country). In his opening address, Boutroux announced, "Since our Congress thereby has its place marked among the Congresses of the present Exposition, a festival of spirit ennobling matter, it could not fail to appear to all as so excellently appropriate given the current state of philosophy."[42] The discipline's evolving relationship to the human and life sciences was paramount among Boutroux's interests. "The theory of knowledge has at this point in time been deepened following this double spirit at once scientific and philosophical. We are no longer content to analyze the data of knowledge or to seek dialectically for the conditions of science in general. . . . We have taken as our subject matter the sciences as they exist."[43] Between the lines read a critique of the unity of science. No longer only physics and geometry, but also psychology and physiology cast light on the material world. More importantly, Boutroux's words lent support to the symbolism of the event. The International Congress of Philosophy enshrined the *Review of Metaphysics and Morals* as the institutional mouthpiece of French philosophy.

The International Congress of Philosophy took place alongside numerous other society conventions in Paris, which met intermittently during the six months of the exposition. Each represented distinct sciences vying for legitimacy on the European stage. For example, the International Physical Congress first convened that year alongside the fourth meeting of the International Congress of Psychology, which had begun in 1889.[44] Drawing from diverse national communities, the meetings disseminated knowledge beyond the confines of state boundaries. Professionals gathered to debate and determine the methods, standards, and definitions of their disciplines. The International Congress of Philosophy was no different. Philosophers solicited their recognition as a sphere of rigorous inquiry alongside the sciences. Judged by its longevity, the congress (now called the World Congress of Philosophy) may very well be considered a success. Since the inaugural meeting in 1900, it has continued to convene every four years.

Congress sessions were patterned after the French academic philosophy program.[45] They were dedicated to metaphysics, politics and sociology, as well as the history of philosophy. One panel in particular, titled "Logic and History of the Sciences," drew wide interest.[46] The two subjects reflected the organizers' neo-Kantian convictions: that the languages of logic and the history of science were enunciated in a universal register that transcended national vernaculars. The head of the section, Louis Couturat (1868–1914), sought to do just that. He promoted Esperanto, the international language invented in the late nineteenth century out of French, Italian, and Slavic elements. The new lingua franca was conceived to replace bellicose national identities with a global community, and thus to lay the conditions for perpetual peace.[47] Eventually frustrated by Esperanto's imprecision, Couturat went on to invent Ido.[48] The alternative language was designed to be more rigorous. It minimized ambiguity by ensuring that each word (or sign) corresponded to a single idea (or signified)—what Couturat called the principle of univocity.[49] The conviction that science and internationalism went hand in hand animated the congress proceedings. Reflecting on this "intellectual and philosophical cooperation," Brunschvicg remarked that the congress prefigured the League of Nations.[50] The organizers were part of a wide network of intellectual associations, which offered a template for diplomats seeking to institutionalize "international peace" following the wreckage of the Great War. However, tensions were inescapable. Just as the League of Nations eventually broke apart in the face of the nationalist movements that precipitated World War II, the International Congress of Philosophy also failed to remain truly international.[51] The majority of communication was in French. Participants from America, Britain, Germany, and Italy were simply expected to adapt.

Propelled by the momentum of the International Congress of Philosophy, the organizers put in place the French Society of Philosophy (Société française de philosophie) in 1901. Its goal was to advance the same spirit of intellectual exchange, but for a restricted membership of forty-four professors. In his opening remarks to the society, Léon presented the reasons for such an exclusive gathering, "that in philosophy more than anywhere else we would fear, in opening a society such as this to the public, easy or even sometimes extravagant discussions."[52] Chief among the society's tasks was the creation of a philosophical dictionary. André Lalande (1867–1963), a lycée professor who served as secretary general, directed the initiative. He believed that for philosophy to be rigorous, its practitioners ought to agree on a technical vocabulary.[53] The discipline would thus achieve the status of what Lalande called a "normative science" alongside the natural sciences.[54] Many of the society's meetings were dedicated to discussions about the meanings of key terms, as well as their cognates in German, English, and Italian. The initiative continued for over twenty years, as members debated concepts from *Âme* to *Zététique*.[55]

The project was not without its critics. Bergson questioned, "Is it useful to halt terms definitively when the ideas are still flowing?"[56] To his mind, philosophy was a fluid practice whose working concepts shifted as new questions emerged. "In metaphysics and morals, there are words which are and must only be, for the moment, statements of problems."[57] Nevertheless, the aim of the dictionary was not only terminological accuracy; it was also social efficacy. The society aspired to carry out the political vision of the *Review of Metaphysics and Morals* and to steer the cultural tutelage of the country's youth. To that end, members debated over terms' place in the state philosophy curriculum. Lalande worried, "Philosophy students are often disoriented if they move from one class to another. . . . Where there are only nuances, they believe they're seeing only irreducible differences. Hence the skepticism about ideas: everything can be said, everything is taught, everything can be proved."[58] When the first edition of the complete two-volume dictionary was released in 1926, it balanced Bergson's and Lalande's competing goals. The title, *Technical and Critical Vocabulary of Philosophy* (*Vocabulaire technique et critique de la philosophie*), gestured to its organization: entries included a technical definition, which reflected the society's collective agreement, as well as critical discussions of their meanings.[59]

The dictionary was one of numerous projects in the early twentieth century to advance the epistemological convictions of the neo-Kantian coterie that had originally gathered around the *Review of Metaphysics and Morals*. For these thinkers, neither the powers of the intellect nor the concepts of reason were given for all time. Both fueled "the idea of critique," which, according to Brunschvicg, broadened "the truth of the transcendental method."[60] The push and pull between conceptual reflection and social reality would cumulatively advance humans' scientific knowledge and moral sensibility.

Intellection and Intuition in the Institutions of Philosophical Exchange

It was in these institutions—congress sessions, society meetings, lecture courses, and reading communities—where disputes over philosophy's relations to the sciences overlapped with contestations over the nature of selfhood and spirit. Does spirit primarily consist of intellectual reasoning or embodied experience? Does scientific inquiry hinge on testing conceptual systems or on practical negotiations with material reality? At the center was the neo-Kantian picture of the self advanced by Halévy, Léon, and—above all—Brunschvicg: that knowledge expresses the active contributions of the intellect. And in turn, the intellect lends rigor to scientific explanations. It's important to note that these thinkers did not consider the intellect in psychological terms. Intellection constituted a social and historical undertaking. Philosophers' prerogative, they urged, was to reflect on and refine this undertaking. For Brunschvicg, "philosophy is essentially organizing: it transfigures nature from which it discovers not a fixed substance and a mass of facts, but a system of ideas and a strictness of thought."[61] Herein lay neo-Kantians' idealism: philosophy and science occupied distinct yet complementary arenas. Whereas scientists collected the empirical data provided by nature, philosophers perfected the logical activity definitive of human freedom.

In response, Bergson, Fouillée, and Le Roy argued that the intellect offered one—but by no means the only—basis for an epistemology of science and selfhood. They took the intellect to be a tool that facilitates interactions with one's environment. It cuts apart experience. But the intellect fails to comprehend its own limits. This corporeal pragmatism drove spiritualist thinkers' critique of neo-Kantianism. Metaphysics, they argued, shines a light beyond the intellect's limits. These thinkers traced the intellect's origins in embodiment. From this perspective, science and philosophy—although distinct concepts under the intellect's purview—would be shown to emerge out of the same pragmatic interests of their creators.

The neo-Kantian rejoinder was steadfast: spiritualist thinkers embraced "spontaneous consciousness," a false picture of selfhood premised on the aleatory flux of personal experience.[62] Their reliance on individual psychology and the aesthetics of creativity amounted to what Brunschvicg called a "subjectivism without reserve."[63] In his eyes, Bergson's work stood out for framing knowledge—and above all, scientific knowledge—as a figment of corporeal life, bereft of the intellect's reasoning power.[64]

Brunschvicg developed his stance in the classroom. In a 1903 philosophy course at Lycée Henri IV, he presented a lectured titled "The Notion of Spirit in Contemporary Philosophy," in which he asked the students, "Are there not today two manners to understand the affirmations of spiritualist philosophers?"[65] On the one hand, there was a "dynamist" movement that had enshrined volition as the locus of selfhood. On the other hand, Brunschvicg proposed a "rationalist" current. At stake were rival concepts of spirit. We can see in the notes written by a student in the course that Brunschvicg spent the lecture tracing the opposition across what he called the "six fundamental theses of spiritualism."[66] His analysis of two varieties of spiritualism ([a] dynamist; [b] rationalist) was so thoroughgoing that it deserves to be cited at length.[67]

1.	Specificity of psychological facts in regard to physiological facts	
	a. The soul is a reality like other parts of the body.	b. Idealism
2.	Theory of the faculties	
	a. Originality and specificity of faculties	b. The rationalist thesis which gives unity to the activity of the soul
3.	Liberty	
	a. Liberty is the negation of determinism; it is contingency, the originality, of our being, novelty.	b. Liberty is founded on reason, on responsibility. . . . Man is all the more free as he becomes more intelligible to himself.
4.	The existence of the self [*moi*]	
	a. Each is himself by the past; by memories, all of the past is particularized in each of us.	b. Rationalist conception of the self
5.	What is reason in us?	
	a. There is a tendency to arrest the principles of reason since intelligence is something essentially transformable.	b. To the contrary, there is in us an activity of legislation.

6. Supersensible reason: the objective value of knowledge
   a. Being is not knowledge; it's activity. It is by activity (Maine de Biran) that we come in contact with others. And perception is action (Bergson).
   b. Conversely, one could wonder how we could make *science.*

It's striking how cleanly Brunschvicg distinguished the competing doctrines within ideas of spirit. Although his elaboration of the "rationalist" current in column b was sparse, the student's notes illustrate how his professor sought to advance a new curriculum and thereby take over the mantle of spirit from the official state program.

To better comprehend what this rationalist current amounted to, we should look to the published articles in *Review of Metaphysics and Morals.* Therein, the founders took aim at spiritualist thought for having undermined the Kantian enterprise. Kant held that autonomy is impervious to causal analysis; his transcendental method foreclosed the possibility of uncovering the material bases of consciousness. The spiritualist claim that science and philosophy found their genesis in embodiment shook the division of labor safeguarding the realm of freedom from that of nature. To the contrary, the founders argued that our conceptual activity could not be caused by anything found in matter since the very concept of causality with which we understand matter issued from the intellect. Against Kant, however, French neo-Kantians believed that the table of conceptual categories was too rigid, at least as it had been outlined in the first *Critique.*[68] The sciences had shifted. New domains, including psychology and sociology, made it evident that scientific models were not settled. Halévy and Brunschvicg argued, "Scientific laws are neither necessary forms, nor necessary moments of thought; without contradiction they would be able to be other; they are not any more pure fictions, while they succeed; they are inventions thanks to which intellectual activity assimilates things; scientific laws are, in the last analysis, the habits which form liberty."[69] This rationalist conception of freedom was, in the eyes of the journal's founders, the true identity of spirit.

Fouillée saw things differently. He forcefully articulated a central line of spiritualist thought: that the human and life sciences had rendered the division between nature and freedom porous. "I know it has become fashionable among some thinkers, thanks to the excessive propaganda of Kantianism and especially French neo-Kantianism," he wrote, "to transpose philosophy outside of science, under the pretext of placing it higher and making a place for belief."[70] His point was that philosophers' jurisdiction was not confined to the foundations of scientific reasoning. Philosophy ought also to appropriate scientists' findings in order to update outmoded views of the self. What the founders of the *Review of Metaphysics and Morals* found so disconcerting, however, was that spiritualist thought would go so far as to question the whole of the Kantian project. This was built on three interlocking premises:

1. The conceptual division of labor between nature and freedom served to safeguard autonomy by separating the laws determining human representations of the world from the laws regulating human conduct.
2. Knowledge depends on the dual powers of receptivity and spontaneity. Consciousness receives the content of experience by means of intuition and spontaneously unifies that content as an object by means of concepts. Our representations are thus self-conscious, predicated on the capacities to sense and to determine the logical relations of thought.
3. Metaphysics fails to attain knowledge of the self, whereby "metaphysics" is understood as a method of inquiry on the basis of thought alone, devoid of sensory experience. Knowledge of the self is thereby limited to the intellectual activity of applying concepts.

In their overlapping claims to bridge the realms of nature and freedom, the spiritualist critique of the Kantian architecture above converged around a threefold argument that knowledge is not concerned primarily with representations, but with action; that applying concepts is one—yet not the sole—activity of spirit; that metaphysics is not only possible but also necessitated by the problems confronting the human and life sciences. Each in their own way, these thinkers showed how scientific literature offered a *point d'appui* for philosophy to generate an elevated understanding of selfhood at the interstice of nature and freedom.

Bergson did just that at an early meeting of the French Society of Philosophy. He enjoined his peers no longer to take their picture of matter from physics and geometry. The new sciences of physiology, psychology, and neurology offered a more supple, multifaceted picture of our bodies' corporeal circuitry. His philosophical challenge threw a wrench in the Kantian apparatus: if consciousness interpenetrated the nervous system, then how might spirit make use of the body's materiality? Bergson called this project "positive metaphysics." He elaborated, "A metaphysics that begins by molding itself around the contour of such facts [connecting spirit and matter] would have many of the characteristics of an undisputed science."[71] Bergson's method was continuous with his earlier work *Time and Free Will* and *Matter and Memory*, in which he expanded the bounds of positive facts to include the data of immediate experience. "Positive metaphysics," he argued, developed in lockstep with the sciences: "Let us work to grasp experience from as close as we can. Let us accept science with its current complexity, and let us recommence, with this new science as our raw material, an analogous effort to that which the old metaphysicians carried out on a much simpler science. We need to break out of mathematical frameworks, to take account of the biological, psychological, and sociological sciences, and on this broader base construct a metaphysics that can go higher and higher through the continual, progressive, and organized effort of all philosophers, in the same respect for experience."[72]

The decisive point of Bergson's talk—what made it so pivotal in the "neo-materialism" animating spiritualist thought—was his elaboration of a critical method that rivaled Kantian critique. Kant had developed the transcendental method to derive the necessary conditions of possibility (*Bedingungen der Möglichkeit*) for knowledge. He thereby demonstrated that logical rules apply with "apodictic" certainty to the objects of scientific inquiry. Conversely, for Bergson, the aporias of science furnished the conditions with which to generate new conceptual possibilities. Philosophy would not derive conditions of possibility for knowledge, but identify conditions for surpassing the bounds of knowledge. Specifically, neurology and psychology revealed the material limits of thought, to which metaphysics committed what may be called an act of spiritual violence: to transcend those very limits "because the increasingly precise determination of the relation of consciousness to its material conditions, by showing us with growing accuracy on what points, and in which directions, and by which necessities our thought is limited, would thus guide us in the very special effort that we have to make to free ourselves from this limitation."[73] In place of the transcendental method, Bergson posited a genetic method.[74] Its aim was to trace the genesis of our concepts, such that they might be redirected to generate higher modes of selfhood—in other words, richer vectors of experience. By departing from the positive, metaphysics could thus expand the reach of freedom into nature.

"Spiritualist positivism" was the name that Le Roy gave his project. His primary concern was neo-Kantians' commitment to a representational picture of knowledge, according to which thought consists of mental representations (*Vorstellung*). These are the conceptual frames in which the mind, as it were, packages experience. Instead, Le Roy was a pragmatist. He argued that thinking primarily relies on corporeal actions. "In short, the new philosophy is neither . . . a philosophy of emotions nor a philosophy of the will: it would instead be a philosophy of action. Now action," Le Roy clarified, "without a doubt implies emotions and the will: but it also implies something else, in particular, reason."[75] He made sure that spiritualism not lose sight of rational thought, whose ultimate purpose was to facilitate the self's negotiation of a shared biological and social world.

Le Roy developed his pragmatic epistemology after having studied mathematics with Henri Poincaré (1854–1912) at the École normale supérieure. At an early age, Le Roy's sights were set on the university. Born in 1870, he was raised in a sailing family in the coastal town of Le Havre. Le Roy entered the École normale supérieure in 1892 and earned his doctorate in mathematics by 1898. During that time, Poincaré penned a series of articles in which he argued that scientific propositions do not apply to natural phenomena absolutely, but instead constitute useful conventions to facilitate experimentation.[76] The articles were an intellectual stimulant for Le Roy. He came to see that the test of propositions' validity turned on their value for organizing and manipulating facts. Above all, propositions were useful in respect to a comparative field of competing propositions. The other stimulant was a stone's throw from the École normale. At the Collège de France, Le Roy attended Bergson's public lectures. His thought held the key, so Le Roy believed, to excavate the metaphysical consequences of the mathematical revolution in non-Euclidean geometry. Reflecting on Le Roy's life work, the philosopher Étienne Gilson wrote, "The most apparently 'Bergsonian' of those who carried the mark of his influence, Édouard Le Roy had always championed teaching the same doctrine as that of the philosopher to whom he had offered such ardent homage."[77] Such portrayals have cast him as a derivative thinker whose writings merely tinkered with the more profound work of Bergson and Poincaré. Yet, Le Roy made timely and incisive interventions in the disputes between spiritualist and neo-Kantian thought.

In a series of articles published in the *Review of Metaphysics and Morals* titled "Science and Philosophy," Le Roy construed the titular concepts as divergent practices sharing a common source in the self. He outlined the genesis of scientific and philosophical thought along three strata: first, embodied engagement with the environment; second, "discursive reasoning," which renders practical activity intelligible by means of symbols; third, creative speculation, the mind's departure from the body's practical exigencies. Spiritual life thus progressively emerged from corporeal experience. Along the way, the body's action-oriented perception cut the contours perceived in the material environment, making it so "the border of an object marks only the limit where an object ceases to interest our action and, since we act principally by contact, it is most often tactile and muscular impressions that determine its borders."[78] Here, Le Roy's pragmatism informed a phenomenology of embodied perception. He described the initial division (*morceler*) of continuous space into discrete entities, which discursive reasoning in turn elevated to a heightened degree of abstraction. These formed the working categories of the sciences. By setting the intellect's powers in continuity with the body, Le Roy demonstrated that scientific symbols were not simply arbitrary. Their emergence owes to the body's demands. The upshot was that Le Roy assaulted the neo-Kantian view that concept formation constitutes a *synthetic* activity, reflecting the intellect's judgments and the sensory data of intuition. Le Roy took perception to be *subtractive*. The sciences arrive at generalized principles and reproducible results by fragmenting the concrete fluidity of experience. In turn, philosophy would elevate higher. It ceases to be practical and speculatively reconstructs the unity torn asunder by science.

Fouillée articulated a similar epistemology of science and selfhood. It was based on his philosophy of *idées-forces*, according to which ideas do not constitute mental representations but forms of action. He saw scientific reasoning as one such form of action whose aim was to subtract the content of immediate experience. Fouillée argued, "The first knowledge [of science] divides the indivisible, homogenizes the heterogeneous, immobilizes the mobile, renders static the dynamic."[79] The sciences depend on the body's transformation of the material world into manipulable units. This is the process that Fouillée described in his tome *The Psychology of Idea-Forces* (*La psychologie des idées-forces*) (1893), as "the primitive and fundamental classification of ideas according to their relation to our volition."[80] The categories with which the sciences quantify, organize, and explain empirical data ultimately serve the human organism's activity. From this perspective, Fouillée wrote: "All our images end up being spontaneously arranged and classified as in a sphere whose center we occupy and whose circumference appears to dilate or focus all around us. Thus, in a flow, objects come to take their place more or less far away according to their diverse densities, some rising like a cork toward the surface, others falling like lead toward the depths; it is the symbol of order established by itself between our representations, not only according to their respective strengths or qualities, but also according to their relation to our central volition."[81] Scientific symbols thus function as tools for organizing matter. And the intellect is one of many malleable devices that help humans do just that.

In response, the neo-Kantian criticism of Bergson, Fouillée, and Le Roy reasserted the primacy of the intellect. Brunschvicg argued that each thinker clung to a jaundiced view of the intellect, one that took discursive reasoning to be monolithic and static. Against this foil was juxtaposed the dynamic continuity of embodied experience. Brunschvicg insisted that the intellect need not be so rigid. Intellection comprised more than logic chopping. At its root was judgment. He claimed that the intellect's tools (i.e., concepts and reasoning) rely on judgments. Concepts link properties to individuals. Reasoning negotiates these linkages. Resultantly, acts of judgment use both to affirm that something is the case. In his book *The Modality of Judgment* (*La modalité du jugement*) (1897), and again in *Introduction to the Life of Spirit* (*Introduction à la vie de l'esprit*) (1900), Brunschvicg argued that judgments consist of two kinds.[82] They can be of internal relations: such as the judgment that the interior angles of a triangle are equivalent to two right angles. The relations are internal since they inhere in thought. Judgments can also be of external relations: that dinosaur fossils, for example, are indicative of life on earth before humans. It is a judgment of a fact given in experience; it exists in the world as it appears to us. The two modalities of judgment turned on what could be arrived at by inner reflection and what could be known owing to the shock of reality. Most judgments combine a mixture of these modalities. But neither, Brunschvicg claimed, is reducible to the body's practical demands.

Brunschvicg defended "the real character of intellectualism" in a critique of Le Roy.[83] The philosopher-mathematician had disfigured the intellect by "deliberately breaking up the synthetic unity of thought and returning it to the static elements on which the dogmatism of long ago was founded."[84] As Brunschvicg saw it, Le Roy drove a wedge between the intellectual and immediate aspects of experience. Judicious writing was not Le Roy's knack. For example, his claim that "science is a ruse of the spirit for conquering the world"[85] invited misgivings. He often lent the impression that science was invariable, whereas metaphysics was the lone creative mode of thought. Yet, Le Roy took care in his response to Brunschvicg to clarify that the body's motility was better attuned to the requirements of science than were judgments. The former consists of "the entire ensemble of preliminary movements by which we prepare ourselves to seize an object, describe its contours, experiment with its functions, palpate it, move it, handle it, in sum, to practice with it and live it."[86] Between Le Roy's spiritualist positivism and Brunschvicg's neo-Kantian idealism lay an abyss separating rival visions of science, spirit, and selfhood.

Bergson intervened in the debate in 1903. He wrote an article for *Review of Metaphysics and Morals* titled "Introduction to Metaphysics," which was read as a scandalous manifesto by the journal's founders. Bergson's goal was to distinguish the methods of metaphysics and science in an effort to build up each on the other. The latter makes use of symbols to analyze its objects of inquiry. "Even the most concrete of the sciences of nature, the sciences of life," Bergson wrote, "confine themselves to the visible form of living beings, their organs, their anatomical elements. They compare these forms with one another, reduce the more complex to the more simple, in fact they study the functioning of life in what is, so to speak, its visual symbol."[87] The result is relative knowledge. With Brunschvicg's account of judgment in mind, Bergson claimed this form of knowledge links a general concept with particular content. On Bergson's view, concepts are defined by their attributes of generality and abstraction. The intellect generalizes what is singular in order to compare it to an abstract standard (e.g., length, weight, species, etc.).[88] By contrast, Bergson contended that intuition offers a metaphysical kind of knowledge suited to comprehend objects in their singularity. Intuition, he claimed, is "the *sympathy* by which one is transported into the interior of an object in order to coincide with what there is unique and consequently inexpressible in it."[89] Here Bergson seemed to claim that intuition touches the ineffable—in religious terms, a disclosure of the incomprehensible divinity at the core of experience. His point, rather, was that the intuitive method is not fit to just any object. It is uniquely disposed to comprehend the self. As was the case throughout his oeuvre, Bergson identified the spiritual core of selfhood with volition. By fastening one's attention on this energetic activity, intuition opens a window onto the self—just as in *Time and Free Will* Bergson claimed that one knows lived duration by a "strenuous effort of reflection."[90] In "Introduction to Metaphysics," he continued to emphasize "the essentially active character of metaphysical intuition"—an exertion to break the habits of thought, to cease *applying* concepts, and instead to *invent* new concepts.[91] This involves a rhetorical labor of description; a rich account of experience that pushes language far enough to break out of clichés. One "cuts for the object a concept appropriate to the object alone, a concept one can barely say is still a concept, since it applies only to that thing."[92] Far from having cast doubt on the utility of concepts, Bergson insisted that they lay the necessary conditions for intuition in that their hallmarks—generality and abstraction—offer points of resistance. Here, he went beyond his prior work. With a rigor that rivaled that of his interlocutors in the *Review of Metaphysics and Morals,* Bergson developed intuition into an epistemological method. And he deployed that method via evocative descriptions of immediate experience.

The upshot of Bergson's essay was that intuition had much to offer intellection. Behind the genesis of every new scientific system,[93] he suggested, is an arduous effort of intuition. When problems are posed of the material world, new concepts ensue. The process of concept creation depends on an exertive act to violate the working categories of prior systems. "Science and metaphysics then meet in intuition," Bergson concluded. "Its results would be to re-establish the continuity between the intuitions which the various positive sciences have obtained at intervals in the course of their history, and which they have obtained by a strokes of genius."[94] Although scientists are normally engaged in the rote business of applying concepts to add new data points to their compendia of information (achieving greater predictability, precision, and statistical significance), inventive moments break with such quotidian tasks. In an ode to Le Roy, Bergson added, "Exempt from the obligation of arriving at results useful from a practical standpoint, [intuition] will indefinitely enlarge the domain of its investigations."[95] Animating new sciences are creative selves, people whose efforts surpass the utilitarian bent of intellection.

The implication was not insignificant. In pushing beyond the intellect, intuition attains not relative, but absolute knowledge of the self. Bergson offered an introduction to *metaphysics* in the sense that Kant had used the term: to know the thing in itself (*Ding an sich*). This was a surprising approach. Neo-Kantians had parted with Kant's noumenon. They argued that the idea was superfluous at best and incoherent at worst since all knowledge is relative to our concepts. According to Brunschvicg, "a thing outside of knowledge would be by definition inaccessible, indeterminable, that is to say equivalent for us to nothing."[96] Paradoxically, Bergson mobilized Kant against neo-Kantianism. Yet, his spiritualist reinterpretation of metaphysics came with a twist. He argued that the absolute (*pace* Kant) is not anterior to experience. Rather, Bergson identified metaphysics with experience itself: the incessant modulation of embodied volition, which drives toward the future and, therefore, novelty. In this specific sense, Bergson revived the Kantian noumenon against neo-Kantian phenomena.

One critic saw in Bergson's thought the seeds of a new critical philosophy. "Since Kant, the critique has taken a half-turn. Kant assured the objectivity of science by renouncing knowledge of absolute reality. The New Critique sacrifices the objectivity of science to authorize metaphysics access to the noumenon."[97] For others, there was nothing metaphysical about Bergson's intuitive method. Fouillée argued that it was hardly new: "In short, as an apprehension of *reality,* intuition boils down to the awareness of our existence, the *cogito-sum* of Descartes; as apprehension of heterogeneous *quality,* it boils down to multicolored sensations, to the innumerable shades of the inner rainbow, which are due to our way of feeling and refracting."[98] Although Bergson drew on spiritualist resources, his methodology was not embraced by all proponents of spiritualist thought.

Furthermore, Bergsonian intuition upended the architectonic division between receptivity and spontaneity. As Kant had claimed, our concepts reflect limited capacities peculiar to humans' form of life. In order for an object to appear before the mind, one must be affected. If we were infinite creatures, however, and possessed what Kant called "intellectual intuition," then the mind could furnish its own objects. As he wrote in the first *Critique*, the receptive capacity of human sensibility is "derivative (*intuitus derivativus*) rather than original (*intuitus originarius*), and hence is not intellectual intuition."[99] In parallel, if the understanding's spontaneous power were not constrained to applying concepts, then it could generate its own content. But the concepts or our finite being pertain to possible objects, meaning that the understanding contributes only to the objects presented in sensible experience. The understanding is discursive; its role is to make judgments about particular items we encounter. Bergson had none of it. He argued that human knowledge is not relative; an immediate form of knowledge remained possible. "The whole of the *Critique of Pure Reason* leads to establishing the fact that Platonism, illegitimate if Ideas are things, becomes legitimate if ideas are relations, and that the ready-made idea, once thus brought down from heaven to earth, is indeed as Plato wished, the common basis of thought and nature. But the whole *Critique of Pure Reason* rests also upon the postulate that our thought is incapable of anything but Platonizing, that is, of pouring the whole of possible experience into preexisting molds."[100]

It is fruitful to interpret Bergson's account of intuition in the context of his critique of neo-Kantianism. Intuition marked his attempt to salvage the "intellectual intuition" that Kant had reserved for a superhuman being. To Bergson's mind, our familiarity with embodied volition constitutes an original, immediate source of self-knowledge. In the Kantian sense, Bergsonian intuition thus revealed nondiscursive knowledge. More broadly, "intuition" could also be said to have informed Bergson's contribution to spiritualist thought: it represented his own effort to enlarge the scope of spirit beyond the myopic lens of the intellect. He returned to this idea throughout his career. At a 1911 conference, for instance, Bergson described intuition as the simple current at the core of thought: "A philosopher worthy of the name has never said more than a single thing: and even then it is something he has tried to say, rather than actually said."[101]

From Intuition to Evolution

Intuition grew out of Bergson's continuous engagement with neo-Kantian and spiritualist thought. Yet, the concept was hardly straightforward. It underwent alterations and revisions across his writings during the first decade of the twentieth century. Initially, Bergson presented intuition as a method of self-knowledge in his talk before the French Society of Philosophy in 1901. In 1907, Bergson published his opus *Creative Evolution*, in which he expanded intuition into a method for comprehending life. Drawing on evolutionary biology, he argued that the self's spiritual powers had emerged out of the creativity inhering not only in the human species, but in the entire natural world: what he called the *vital impulse* (*élan vital*). To explain the vital impulse—and, ultimately, to harness it for the invention of new concepts—involved an act of intuition.

What accounted for this shift in Bergson's thought? Why did he move on from neurology and psychology, the arenas of his early work, and tackle the science of life? Bergson made clear in the opening pages of *Creative Evolution* that one of his goals was to trace the genesis of the intellect. "Deposited by the evolutionary movement in the course of its way, how can it [the intellect] be applied to the evolutionary movement itself?"[102] As he went on to demonstrate, intellection is a powerful tool used by the species *Homo sapiens* to surpass its instinctive capacity to use its body and to make material implements. Intellectual activity brings the organism outside itself. Yet, because it is exercised primarily to forge relations among objects, the intellect proves ill equipped to grasp the creative forces from which it arose. To make sense of this creativity, one must resist the hypertrophy of intellection by an effort of intuition. Bergson elaborated: "When we put back our being into our will, and our will itself into the impulsion it prolongs, we understand, we feel, that reality is a perpetual growth, a creation pursued without end."[103] This reflexive form of knowledge, by which the self turns thought on the experience of its very efficacy, reveals the natural genesis of the intellect out of nonintellectual resources. By the conclusion of *Creative Evolution*, Bergson had opened an evolutionary vantage point from which neo-Kantian thought appeared to inhabit a mere enclave amid a vast metaphysical expanse.

Equally decisive for the shift in Bergson's thought was his profound curiosity in the problems facing evolutionary biology in the early twentieth century. During the period, the very notion of "evolution" was still inchoate. Charles Darwin (1809–82) wrote about "descent with modification"; Ernst Haeckel (1834–1919) referred to "transmutation."[104] Bergson belonged to an early generation of thinkers who used the term "evolution." At issue, specifically, was the nature of convergent evolution. How could the same organic structure take shape in diverse species? Why did human and nonhuman animals—each with distinct genetic lineages—all nonetheless possess eyes with analogous oculi, retinas, and corneas? How could it be that even mollusks grow a light-sensing apparatus, such as the pigment-spot of clams and scallops? That such widely disparate species all possessed similar visual organs was a compelling problem.[105] When Darwin published *On the Origin of Species* in 1859, he had showed that natural selection works by selecting variations among organisms within the same species. Over time, those differences widen as organisms die off and environments shift. As a result, groups of organisms diverge from each other and give rise to new species. During the ensuing half century, it came to be recognized that divergence leads species to follow separate evolutionary paths. Bergson discerned this to be the crux of the problem, since "the more two lines of evolution diverge, the less probability is there that accidental outer influences or accidental inner variations bring about the construction of the same apparatus upon them, especially if there was no trace of this apparatus at the moment of divergence."[106] He found the eye to be such a captivating case of convergent evolution because the organ is essential to organisms' action on matter. Perception, Bergson reminded readers, is not a representation of the world; it is a tool used to interact with the world. The visual field outlines the contours of potential contact. That such a simple activity relies on an intricate organic structure — one that took hold over millennia and across disparate species — gave rise to a host of philosophical problems.

Two aspects of convergent evolution were problematic. First, Bergson discerned that diverse species' eyes developed continuously in the same direction. The ocular tissues of squid, orangutans, eagles, and humans (for example) all underwent variations, which successively converged, beginning with the optic vessels, retina, lens, and extending through the choroid, sclera, and cornea. These variations grew along the same trajectory, coming together in the fully formed camera-like eye. Second, ocular tissues came to be coordinated despite the fact each part of the eye emerged at different moments in history. According to natural selection, some parts would be selected out if they failed to aid the organism. Darwin was aware that not all variations would immediately help an individual yet nevertheless be useful for the species. He claimed that variations are often too slight to be felt by an organism. So, there was no clear reason why every variation in the eye would be preserved. Consequently, Bergson sought to explain — at a macro scale — the continuity among species' evolutionary trajectories as well as — at a micro scale — the coordination of organs' component tissues.

To Bergson's mind, neither of the major evolutionary theories of the period — mechanist or finalist — successfully answered both aspects of the problem. On the one hand, advocates for mechanism claimed that new species equipped with undeveloped organic structures such as the eye had diverged thanks to the gradual aggregation of random adaptations. In the aftermath of Darwin's work, a central point of dispute was how to account for this aggregating process. William Bateson (1861–1926) and Hugo de Vries (1848–1935) argued that insensible variations did not amass over generations; rather, sudden mutations sparked divergence. Rupture, not continuity, was to thank for the eye's transformation.[107] Not all mechanist thinkers, however, believed that variations were random. Theodor Eimer (1843–98) argued that the environment caused species to undergo adaptive modifications.[108] For instance, lizards' skin pigment harmonized with their milieu. Physical and chemical changes in the organism furnished a natural defense mechanism in the form of camouflage. On the other hand, finalism emphasized organs' functions. So, diverse species' eyes converged in order to support vision. This line of thought hewed to the work of Jean-Baptiste Lamarck (1744–1829), whose early nineteenth-century writings had posited the inheritance of acquired traits as the motor of evolution. Species diverged because of organisms' progressive efforts to exploit their environment. By the end of the century, Edwin Cope (1840–97) had revived finalism. He observed the small eyes of burrowing snakes (*Cylindrophiidae*) whose eyelids disappeared because of lack of use. By contrast, snakes living above ground retained their eyelids because they relied on vision.[109] Eye structures converged when species made overlapping efforts to take advantage of light, a crucial source of energy in the environment. That the eye also came to exhibit a coordinated anatomy followed from the limited physiological means available to the species given their structural constraints.[110] But such finalist theories soon lost ground. The rise of genetics (a field still in its infancy at the time) explained species mutations according to what August Weismann (1834–1914) called the "germ plasm."[111] It was just as likely that the burrowing snakes retreated underground because of the degenerate eyes passed down from the parents' sperm and egg cells.

As Bergson saw them, both theories—whether they emphasized the composition or the function of the eye—stubbed their toe against the same stone. Mechanism and finalism assumed that evolution *accumulated* variations over time. For each, "nature has worked like a human being by bringing the parts together."[112] According to Bergson, nature appeared as such only from the perspective of the intellect, which figures evolution's operations according to discursive reasoning. Numerically discrete variations came together over time and brought about species' divergence. Bergson's core argument was "Life does not proceed by the association and addition of elements, but by dissociation and division."[113] The claim repeated a conceptual motif of his prior work. Bergson argued that the continuity and coordination among organs had issued from a single creative activity, an activity whose powers remained obscure so long as they were framed in conceptual pictures borrowed from the Kantian tradition.

Bergson called this creative activity the vital impulse. Of a piece with spiritualist thought, his argument was that its operations were subtractive. "Nature's simple act has divided itself automatically into an infinity of elements which are then found to be coordinated to one idea."[114] Division thus explained convergence. In the case of the eye, "according as [the vital impulse] goes further and further in the direction of vision, it gives the simple pigmentary masses of a lower organism, or the rudimentary eye of a Serpula, or the slightly differentiated eye of the Alciope, or the marvelously perfected eye of the bird; but all these organs, unequal as is their complexity, necessarily present an equal coordination."[115] An identical impulsion to act on matter embodied analogous ocular structures across the spectrum of life.

The vital impulse spurred a host of interpretive difficulties. The French *élan* is not necessary impulsive (which is to say, sudden); the word could better be translated as "momentum" or "vigor," either of which denotes an ongoing energetic process. More importantly, readers at the time were liable to misconstrue the concept with similar evolutionary theories: finalism and vitalism. In regards to the former, it seemed that the vital impulse imparted a purpose to species' evolutionary trajectory. This was especially the case when Bergson wrote in the above passage that the idea "goes further and further" in certain species. He took great care, however, to clarify that the vital impulse did not predetermine organic structures' variations. Bergson eschewed the Aristotelian notion that nature comprises a hierarchy of genera, whereby vegetable, animal, and human life constitute ascending rungs along the same biological ladder. The thrust of his claim was, instead, that species differ according to their complexity. "The will of an animal is the more effective and the more intense, the greater the number of the mechanisms it can choose from, the more complicated the switchboard on which all the motor paths cross, or, in other words, the more developed its brain."[116] The vital impulse therefore "goes further and further" in species with a wider breadth of freedom—that is, in species whose ocular structures serve a nervous system equipped with multifaceted tools for acting on matter. "Vision will be found," Bergson added, "in different degrees in the most diverse animals, and it will appear in the same complexity of structure wherever it has reached the same degree of intensity."[117] This is also what distinguished the vital impulse from vitalism, the evolutionary theory that posited an energetic substance inhering in living organisms and separating them from inanimate matter. Bergson did not construe vitality as a substance. Far from inhering in organisms, the vital impulse offered a framework to make sense of the interaction between action and anatomy. He wrote, "the 'vital' order, which is essentially creation, is manifested to us less in its essence than in some of its accidents, those which *imitate* the physical and geometrical order."[118] Spiritual action on matter was the fulcrum of the vital impulse. It was a product of metaphysical speculation meant to explain what biology could not.

The view that life grows via diversity and complexity (and not via accumulation) hewed to the work of Jean-Marie Guyau. His book *A Sketch of a Morality Independent of Obligation or Sanction* (*Esquisse d'une morale sans obligation ni sanction*) (1885) was ostensibly a work of moral philosophy. But Guyau's goal was to situate morality in a theory of life. More powerful than society's codes of right and wrong were the biological tendencies driving people to empathize with others. Guyau suggested that duty should be understood "not as a limitation or restriction of activity, but as the very consequence of its expansion."[119] Life involves a dialectic between expansion, which drives humans toward sociality, and contraction, which engenders private egoism: "life has two aspects. According to the one, it is nutrition and assimilation; according to the other, production and fecundity. The more it takes in, the more it needs to give out; that is its law. Expenditure is not an evil physiologically; it is one of the conditions of life."[120] For one to cultivate expansive sentiments, Guyau argued it is imperative to carve out an inner space to experience, reflect on, and activate not only a fidelity to the social good. "The moral ideal will be *activity* in all its *variety of manifestations*."[121] This ethics of the self encouraged the diversification of one's pursuits: to eat new foods, to visit other cultures, to speak different languages, to read multitudes of books. Guyau feared that a moral philosophy predicated on the intellect alone—and not on instinctual urges—risked stifling our vitality. Exemplar was Kant's categorical imperative, which derived a universal rule for duty. "In reality, reason in the abstract is incapable of explaining a power, an instinct, of accounting for a force which is infra-rational in its very principle. Observation, experience, is necessary."[122] Whether Guyau's book had a direct influence on Bergson's is unclear.[123] It did, nevertheless, offer a formidable contribution to spiritualist thought, which was to show how biological growth engenders greater power to organisms in the form of heterogeneity and diversification.

Bergson went further than his fellow spiritualist in tackling the conceptual problems *within* biology. How are species' organic structures entangled with their capacities for action? What are the natural linkages connecting physiology and volition? How do material constraints enable conceptual creation? Bergson's answer was not only that the vital impulse motivates species to develop diverse organs, but that in so doing species also pose problems of their material environment. Energy is one such problem. Species diverge in respect to the anatomical solutions created to exploit natural resources. And they converge when their anatomies yield shared solutions to the problem—the eye being one such solution. Bergson wrote: "The role of life is to insert some *indetermination* into matter. Indeterminate, i.e. unforeseeable, are the forms it creates in the course of its evolution. More and more indeterminate also, more and more free, is the activity to which their forms serve as the vehicle. A nervous system, with neurones [*sic*] placed end to end in such wise that, at the extremity of each, manifold ways open in which manifold questions present themselves, is a veritable *reservoir of indetermination*."[124] Complexity, novelty, and indeterminacy go hand in hand. As Bergson saw it, life is creative to the extent that new variations enable species to use the environment in unforeseen ways. He succinctly articulated the point in a 1911 lecture: "This is precisely what life is—freedom inserting itself in necessity, turning it to its profit."[125] Among the tools uniquely at the disposal of human organisms, the intellect perfects our manipulation of material objects. Yet, this tool is too blunt, Bergson claimed, to comprehend the workings of creativity within nature.

How could it be that intellection would get life so wrong? If discursive reasoning, conceptual analysis, and numerical aggregation were all higher-order mechanisms that had evolved in *Homo sapiens* to take advantage of material reality, then why would they confound the species' attempts to comprehend their very genesis?

Brunschvicg seized on this point. His response to *Creative Evolution* charged that Bergson had lost sight of the intellect's true powers. Life's creativity, Brunschvicg argued, meant precious little apart from the intellectual capacity to subsume the instances of evolution under explanatory concepts. The vital impulse represented, on his account, such an achievement of intellection. And its conceptual power could be made all the more rigorous thanks to math and physics, which by the twentieth century had become "a supple and living instrument, infinitely plastic and infinitely fruitful, intended to capture and to make present—if not to the senses, at least to the intelligence—qualities that escape the infirmity of our organism and perception."[126] In Brunschvicg's eyes, the intellect was evolution's finest achievement because it endowed humans with the capacity to make sense of the very creativity within evolution.

Although the intellect could explain many things, Bergson claimed that its own genesis was not among them. Like all organic tools, the intellect had diverged from the ensemble of other tools that had evolved in humans. Because evolution is ongoing and never complete, the vital impulse would slip through the intellect's grip. It followed that biology could analyze life's creations (*natura naturata*) but not its constitutive creativity (*natura naturans*). The latter was left to metaphysics. According to Bergson, one could open a subjective window onto the inner operations of the vital impulse by undergoing an effort of intuition. In resisting the habits of intellection, the self feels the very strain by which life imparts creativity to living beings. "Let us try to see, no longer with the eyes of the intellect alone, which grasps only the already made and which looks from the outside, but with the spirit, I mean with that faculty of seeing which is immanent in the faculty of acting and which springs up, somehow, by the twisting of the will on itself, when action is turned into knowledge, like heat, so to say, into light. To movement, then, everything will be restored, and into movement everything will be resolved."[127]

Here, Bergson characterized intuition as a proto-phenomenology of life. What the self intuits is an immediate contact with the vitality within oneself; a reflexive knowledge, as Bergson put it, of "twisting the will on itself." Although they took biology as their point of departure, the epistemological methods with which to understand the vital impulse are not entirely biological. Spirit's part is decisive.

In peering inward on life's creativity, Bergson returned to the psychological themes that had been the subject of his prior work. "At every instant," he wrote, "evolution must admit of a psychological interpretation which is, from our point of view, the best interpretation."[128] The reasons were numerous. To begin, psychology offered an antidote to the physical models of the natural sciences. "[Life] must be compared to an impetus, because no image borrowed from the physical world can give more nearly the idea of it. But it is only an image. In reality, life is of the psychological order, and it is of the essence of the psychological to enfold a confused plurality of interpenetrating terms."[129] Ultimately, however, life was not only comparable to psychology. Psychological concepts served to address the aporias at the heart of convergent evolution. Bergson explained how the whole of life retained diverse parts, such that the vital tendency to pose problems of matter generated analogous organic structures: "The elements of a tendency are not like objects set beside each other in space and mutually exclusive, but *rather like psychic states,* each of which, although it be itself to begin with, yet partakes of others, and virtually includes in itself the whole personality to which it belongs. There is no real manifestation of life, we said, that does not show us, in a rudimentary or latent state, the characters of other manifestations."[130] In other words, individual anatomical structures such as the eye preserved an organic memory common to all organisms. By presenting convergent evolution as an effect of memory, Bergson drew from his earlier work on duration. In a letter to Le Roy, Bergson confirmed, "None of my books could be deduced from the preceding, like the extension of a straight line. Each of them tries to enlarge the circle of reality that the previous studied; they are concentric attempts; if one wanted to pick out the last from the first, one would come up against the same absurdity as if he tried to pull the container from the contents."[131] The enlargement of psychology into biology was made clear in the early pages of *Creative Evolution,* "Wherever anything lives, there is, open somewhere, a register in which time is being inscribed."[132] Just as time is inventive, so too is evolution creative.

What ensued was a picture of nature very much of its era. Biologists have since moved beyond the debates in which evolutionary theory was embroiled over a century ago. Notably, Bergson and his contemporaries were unaware of the molecular structure of DNA, which James Watson and Francis Crick identified in 1953. Instead, the picture of nature offered up by Bergson was one in which humans inhabited life's zenith. In fact, *Creative Evolution* could be read as a work that enjoined readers to take a creative leap with Bergson. Perhaps his aim was not only to offer new concepts to explain evolution, but also to spur an evolution in thought. What is clear is that Bergson's chief interest was humans. His view was unapologetically anthropocentric. Among the myriad evolutionary pathways, Bergson admitted, "not all of these directions have the same interest for us: what concerns us particularly is the path that leads to man. We shall therefore not lose sight of the fact, in following one direction and another, that our business is to determine the relation of man to the animal kingdom, and the place of the animal kingdom itself in the organized whole as a whole."[133] Species other than *Homo sapiens* remain ignorant of the vital impulse. Limited to their instincts, nonhuman animals are unable to harness the spiritual dynamic at the core of life's creativity. Only in humans could it be fully realized by an act of intuition.[134] In this sense, *Homo sapiens* sits atop the natural world for reasons that are not at all exclusive to the species. What is intuited within oneself is not entirely one's own. As Maurice Merleau-Ponty described Bergson's method: "It is not necessary for him to go outside himself in order to reach the things themselves; he is solicited or haunted by them from within."[135] This anthropocentrism was predicated on a spiritual power that precedes and exceeds humanity.

Looking back on the disputes over science, selfhood, and intellection at the turn of the twentieth century, what comes into view are the conditions (as well as constraints) in which spiritualist thought evolved into a "neo-materialism." This evolution came about thanks to the resurgence of neo-Kantian philosophy among interlocutors who trafficked in conferences, reading communities, educational institutions, and scientific associations.

Across these shared channels, the tension between idealism and spiritualism was prominent and productive.

But that tension also proved to be all encompassing. Despite their stringent claims to overcome the Kantian heritage, Bergson, Fouillée, and Le Roy never entirely escaped the architectonic division between nature and freedom. In reconceptualizing the corporeal continuity between spirit and matter, these thinkers brought the realm of freedom to the threshold of nature. Indeed, *Creative Evolution* was among the most celebrated attempts to do just that. Yet, a conceptual boundary held firm. As much as Bergson showed how the vital impulse animated the living formations in the material world, its spiritual creativity remained impervious to its material forms. Exemplary was Bergson's vivid assertion that "the consciousness of a living being . . . is inseparable from its brain in the sense in which a sharp knife is inseparable from its edge: the brain is the sharp edge by which consciousness cuts into the compact tissue of events, but the brain is no more coextensive with consciousness than the edge is with the knife."[136] In the "neo-materialist" moment, concepts of matter oscillated between two figures: a limit against which spirit exerted its effort or the instrument that executed spirit's directives. In either case, spirit set freedom into motion only from outside the materiality found in nature.

Not an indictment of science or positivism, but a creative appropriation of both drove the vibrancy of spiritualist thought in early twentieth-century France. It was a moment when competing visions of spirit and selfhood vied for legitimacy. Indeed, the radicalization of spiritualist thinkers' "neo-materialism" should be interpreted as a series of convergent efforts to expose the shortcomings of neo-Kantian epistemology in the face of the life and human sciences. Although these disputes left the house that Kant had built severely weathered, its infrastructure, ultimately, remained intact.

✷ CHAPTER 6 ✷

Struggles for Spirit's Catholic Soul

Ideas of spirit were inseparable from their deep Catholic roots. The materialist moment did not take shape only in response to upheavals in the human and life sciences; it also cohered around a revitalized understanding of faith prompted by the cultural circumstances of Roman Catholicism in the early twentieth century. The trenchant secular policies of the Third Republic eroded the church's foothold in France. Laïcité sequestered religion to the private realm; faith was not to trouble the public order. Although the institutional foundations of Christianity were profoundly shaken, theological ideas of selfhood underwent a startling resurgence. Two figures in particular, Maurice Blondel and Édouard Le Roy, negotiated the Catholic commitments of spiritualist thought. They made explicit what had become ever more disconcerting for clerical authorities: that faith could no longer remain impermeable to science. Both thinkers brought the critique of intellection (the subject of the previous chapter) into direct confrontation with the church. Blondel and Le Roy drew on psychology and the life sciences to reexamine faith within the framework of personal experience.

Blondel and Le Roy, the subjects of this chapter, fanned the flames of the modernist crisis within the church. This theological rebellion sought out new ways to conceive of God's truth so that it could become human beings' spiritual possession. In their respective efforts to enlarge the scope of religious experience, Blondel and Le Roy upended the long-standing scholasticism that took Catholicism to constitute a theological system. In response, the Vatican issued severe condemnations of modernism and doubled down on its commitment to reason. To be sure, Blondel and Le Roy did not see eye to eye. To varying degrees, they negotiated the fraught boundaries between contemplative belief and embodied practice. In their sweeping challenges to orthodoxy, they made ideas of spirit matter afresh for religion.

Catholicism had been integral to France's cultural heritage. The church, *Ecclesia Gallicana*, although connected to Rome, had retained its own peculiar privileges and rights since the Concordat of 1516 between King Francis I and Pope Leo X. State power and ecclesiastic authority were intertwined. Yet, in the aftermath of the French Revolution, Catholicism could no longer be taken for granted as a definitive feature of French identity. An enduring fissure separated reason and religion. These were *les deux France*.[1] Throughout the nineteenth century, Catholics voiced royalist aspirations for a return to the monarchical glory of the ancien régime; Republicans espoused the secular and universal imperatives bequeathed from the Enlightenment. The tension reached a climax in 1882, when President Jules Grévy barred religious instruction in primary schools, leaving it up to families and private schools to inculcate Christian teachings. In 1905 parliament announced the official separation of church and state. This period of anticlerical reforms under the Third Republic challenged the church's role in public life and stoked the modernist crisis within Catholicism.

Catholic modernism was a variegated and multinational movement, which erupted among theologians who sought to reconcile the cultures of reason and religion. Psychology eroded Christian doctrine by subjecting the immaterial realm of the soul to laboratory inquiry. Biology exploded traditional visions of time that had endured since the era of the church fathers, premised on the belief that the earth was six thousand years old.[2] In France, Blondel and Le Roy showed that such developments were compatible with Catholic teachings. Each in their unique ways, they seized hold of the modernist crisis as an opportunity to reanimate Catholicism in the face of the Vatican's diminishing legitimacy. For many modernists, it was high time to shield Catholic teachings from scientific scrutiny no longer; they galvanized the vitality of scripture from within science. Rome would have none of it. A defensive crouch among papal authorities came to dominate the politics of the Holy See. The works of alleged modernists were officially banned in 1907. Committed to a censorious traditionalism, Pope Pius X went about expelling proponents of the movement from the clergy.

In this chapter I examine the intimate relationship between French spiritualism and Christianity through the prism of Blondel's and Le Roy's interventions in Catholic modernism. Both were avowed Catholics who pursued science as a route to divinity. Philosophy would thus prepare the discovered truths of the sciences as the point of departure for the revealed truths of religion.

As for Le Roy, he aligned himself with the modernist cause. His theological views developed in close dialogue with spiritualist contemporaries. At a 1902 meeting of the French Society of Philosophy, he boldly argued for a critical and historicist approach to Christianity: "There has been during the course of history an immense and continuous evolution in Christian morality. All the while holding on to the same general direction, it has undergone infiltrations, it has adapted in many respects to diverse conditions, it has gradually had to take account of new facts, it has, in brief, enriched and complicated itself like a living thing. It would be ignorant to only consider it in one of its lone moments."[3] Le Roy's "creative evolution" affronted the church's claim that its edicts were eternal. His views dovetailed with those of the Irish Jesuit George Tyrrell (1861–1909), who described Christianity as "a spiritual organism in whose life we participate."[4]

Blondel, however, would not go so far. Resistant to the biological bent of modernism—a position he criticized for neglecting the divine's transcendence—Blondel's commitments to church traditions were profound. He advanced a rival account of Christian history sensitive at once to the contingencies of time and to the ecstatic nature of human experience that had been a pillar of Catholic thought since the Cappadocian fathers.

Nonetheless, both Le Roy and Blondel tackled the same series of problems: Is the church divinely sanctioned? Or is it a human institution that has transformed with the vicissitudes of history? These questions set the stage for spiritualist materialism to contribute to the modernist controversy. At the level of their disputation, we see how Blondel and Le Roy refined pragmatist accounts of selfhood embedded in the passage of history. Stepping back, what comes into view are the points of commonality—as well as divergence—between French spiritualism and more thoroughgoing theologies of selfhood.[5]

By intervening in debates over historicism—debates that called the very idea of revelation into question—Blondel and Le Roy radicalized the intellectual trajectory animating the turn to matter in ideas of spirit. As I have shown in the preceding chapters, the commitment to the will (and not the intellect) drove philosophers' overlapping conceptions of the corporeal points of contact connecting spirit and matter. Historicism offered an additional epistemological framework. Blondel and Le Roy demonstrated that spiritual activity is pragmatic activity that unfolds over time, and moreover, that such pragmatic activity takes on significance in religious tradition. As I'll explore in the following sections, the problem of dogmas (or *dogmata*: the truths of Catholic doctrine) was particularly salient. Against church doctrine, the modernist crisis brought about a fresh understanding of dogmas as living and mutable guideposts for action. For spiritualist thinkers, secularism and science did give rise to a disenchanted world bereft of God, but an intersubjective world that transformed His meaning, a world imbued with the pragmatic significance of the divine within and among persons.

The modernist controversy occasioned and concealed pragmatist accounts of selfhood. In their efforts to reconcile faith and science, Blondel and Le Roy drew the ire of the Vatican. Blondel first used the term "pragmatism" to characterize his philosophy of action: "I had proposed the name *pragmatisme* in 1888 . . . and I was aware of clearly devising it before I had ever encountered the word, which has since been employed in England, America, Germany and Belgium."[6] Blondel promptly retracted the label to avoid the same fate of Le Roy,[7] who fervently raised the pragmatist banner in the face of Pope Pius X's condemnation.[8] By 1907, Le Roy's works were on the Vatican's index of banned books, while Blondel withdrew the new edition of his own.[9]

The religious contexts of spiritualism also shine a light on the emergence of pragmatist thought beyond the shores of America.[10] Although the school of thought is often taken to be distinctly American, a variant of pragmatism took off in the hexagon well before William James (1842–1910) declared it "a new name for some old ways of thinking."[11] Spiritualists' conviction that the value of ideas turn on what they *allow us to do* responded in large part to the shifting power of the church in French society. By looking beyond American soil, my aim in this chapter is to highlight an alternative intellectual historical itinerary, one driven by the cultural urgency to reconceive knowledge as a principle of action *in* the world (rather than of correspondence *with* the world). Further, the religious context in which pragmatism made inroads among ideas of spirit reveals a philosophical divergence from the epistemological commitments often associated with James, Charles Sanders Peirce (1839–1914), and John Dewey (1859–1952). Blondel and Le Roy did construe their pragmatism as a springboard for intervening in the world. It was also part and parcel of their critique of intellectualism, the Kantian view that knowledge is a matter of judging our representations of reality (what Dewey called the "spectator theory" of knowledge).[12] For Blondel and Le Roy, however, pragmatism opened onto the supernatural.

The materialist turn in spirit drew conceptual resources from its Catholic lineage. Maine de Biran had dedicated his late writings to religion.[13] He posited three lives within the self: the life of sense receives the flux of phenomena; the life of thought reflects on the realities behind appearances; and the life of faith sets dialectical reasoning on stable footing. In subjecting sensory experience to critical reflection, Maine de Biran believed that philosophy had generated metaphysical dualities—between the will and the self, effort and knowledge, physiology and psychology—that found their ultimate resolution in theology. "Religion alone solves the problems that philosophy raises."[14] Similarly, Félix Ravaisson drew on a Catholic archive when he articulated the "law of grace" at the center of his opus *Of Habit*. Victor Cousin promoted a legibly Christian morality in the ethics section of the lycée philosophy curriculum, as I discussed in chapter 3.[15] In the sections that follow, I explore how Blondel and Le Roy prompted a reckoning with this lineage. Their critical accounts of science and pragmatism, and in turn, their disputes over the impact of pragmatism on Catholicism, catapulted the materialist moment in spirit to the center of the religious crises in French society at the turn of the twentieth century.

The Modernist Controversy in France

Although Catholic modernism ignited at the turn of the twentieth century, it grew out of preexisting tensions between reason and religion in French society. Under the ancien régime, the Catholic Church offered an institutional and ideological buttress for the divine right of the Bourbon monarchy to rule over the Kingdom of France. The Revolution of 1789 erupted as a violent effort to overthrow the excessive power of the clergy. Religious orders were quickly resolved. Church property was duly confiscated. By 1790, the National Assembly made priests and bishops take an oath of loyalty to the revolutionary government. Only half agreed. Tens of thousands of priests fled the country; many others practiced clandestinely. Bishop Jean-Baptiste Gobel (1727–94), the head of the Paris diocese (and a quite vocal resistant to the oath), was executed. He was one of 794 priests and 126 nuns who met the guillotine under the bloody Terror of 1793 to 1794.[16] What began as a revolt against church corruption transformed into the radical Jacobin's all-consuming program of de-Christianization. The "divorce" between Roman Catholicism and the French national government was put in place.[17] What ensued was an ongoing tension between Catholics' royalist sympathies and the anticlerical prerogatives of Republicans in nineteenth-century France.

The tension endured despite moments of reconciliation. When Napoléon staked his claim to emperor in 1804, he mounted his legitimacy on the twin pillars of revolutionary government and clerical authority. Such was his imperial persona. Pope Pius VII presided over Napoléon's coronation in the Cathedral of Notre Dame (a ceremony perpetually fulfilled, as Pius VII was kept from returning to Rome until the demise of the Napoleonic Empire in 1814). Significantly, Napoléon had resuscitated the Concordat in 1801, which acknowledged Catholicism as "the religion of the great majority of the French people." Fifty bishops were granted authority over the state church. They were to be elected by the French government, though each had received canonical institution from the Vatican. By decree in 1808, the university system took Catholic precepts as the basis of its teaching. Such remained the case through subsequent regimes that invoked ecclesiastical powers: the Bourbon Restoration (1814–30), the July Monarchy (1830–48), and the Second Empire (1852–70). When Republican resistance succeeded, so too did anticlerical policies. It was a central objective of national reformers under the Third Republic in the late nineteenth century to dissolve the church's authority over French society and selfhood.

Abroad, challenges to the Vatican fomented. Spurred by the collapse of the Papal States and the secular unification of Italy, Pope Pius IX (who had reigned since 1846) summoned the First Vatican Council in 1869. Distraught by his diminishing support, and with an eye to the coming anti-Catholic legislation in Prussia, Pius IX asserted his authority. He reestablished the Vatican's role in Europe with the doctrinal constitution *Pastor Aeternus* ("First Dogmatic Constitution on the Church of Christ"), which enshrined the principle of papal infallibility (in fact, the principle is a modern artifact). Importantly, Pius IX also tackled the rising tide of what he labeled "rationalism" and "liberalism." He doubled down on his *Syllabus of Errors* (1864), which decried the rise of secularism among other heresies.

The list of condemned propositions fueled a conservative renaissance in France. *Ultramontanism* was a cultural movement that sought to amplify the power of the pope beyond the mountains (i.e., the Alps). These *ultras* were inspired by the French priest Félicité de Lamennais (1782–1854). They embraced a robust Gallic piety, an amalgam of royalism and clericalism, which left them rarely satisfied with attempts to reconcile church and state. Under the Second Empire of Napoléon III, a law recognizing religious liberties in education was put in place thanks to Viscount Alfred de Falloux (1811–66). The accreditation of philosophy instructors was suspended in the universities for fear that the discipline promoted the rational values behind the French Revolution. The journalist Louis Veuillot (1813–83) inveighed against the regime, arguing that it had not gone far enough. He championed the *Syllabus* of Pius IX, with its condemnation of "liberty, progress, and modern civilization." The arrival of the Third Republic did not immediately abate ultramontanism. In its early years, the government was led by Patrice MacMahon (1808–93), a general from the Franco-Prussian War for whom religion was a lively dimension of politics. In 1875, construction began for the Basilica of Sacré-Coeur, a political and religious emblem of conservative Catholicism built atop the dead bodies of revolutionaries who hid in the Montmartre hill during the Paris Commune of 1871. As a result, many saw the edifice, which loomed over Paris when it finally opened in 1914, as a provocation to civil war.

The reactionary wave in Rome intensified with the policies of Pope Leo XIII, who sowed the seeds of the modernist crisis when he ascended to the pontificate in 1878. One of his first acts was to write "Aeterni Patris" ("On the Restoration of Christian Philosophy in Catholic Schools in the Spirit (*ad mentem*) of the Angelic Doctor, Saint Thomas Aquinas"), an 1879 encyclical whose primary purpose was to establish the relationship between reason and faith. It drew on medieval methods to revive neo-scholasticism, the doctrine premised on the dialogue between teacher and student by which the latter would logically prepare his mind to receive God's revelation. Manuals were key. Seminary students studied textbooks of Saint Thomas's doctrines to ground their faith in certainty. To this end, Leo XIII was a firm promoter of education; he opened the Vatican archives to church laity and protestants alike. In France, neo-scholasticism was reinforced in ministries, and numerous journals were formed to promote Thomistic philosophy.[18]

Leo XIII's war on modernism was unrelenting. In his 1893 encyclical "Providentissimus Deus" ("On the Study of Holy Scripture"), the pope cast suspicion over the theory of evolution and criticized modernists' efforts to integrate evolution into biblical hermeneutics. He created the Pontifical Biblical Commission in 1902, a body of three cardinals and twelve consultors who advised the pope on matters of scriptural interpretation. In the following years, the commission sent a mushrooming number of works that unsettled doctrinaire neo-scholasticism to the Vatican's index of banned books.

Modernists were united in opposition to what they saw as the arid intellectualism of neo-scholastic thought. Their alternative visions of Christianity, however, cohered only loosely. Broadly, modernism embraced a historicist approach that bridged the patristic sources with the present. Alfred Loisy (1857–1940) ignited the spark with the publication of his *L'évangile et l'église* (*The Gospel and the Church*) (1902). The ordained priest advanced what he took to be a science of Christianity. He argued the Catholic Church had developed since the time of Jesus within and alongside human history. Loisy wrote the book in response to Adolf von Harnack's (1851–1930) *What Is Christianity* (*Das Wesen des Christentums*) (1900), which argued that the essence of the Gospels consists of faith in God the Father. Harnack proposed an analytic method to distill the superfluous elements of Jesus's ministry and arrive at its core. "This method of dismembering a subject," Loisy argued, "does not belong to history, which is a science of observation of the living, not of dissection of the dead."[19] Loisy's historical method did not construe the Gospels as Christ's immediate revelation, but as documents written during the church's early development. Biblical criticism, Loisy held, ought to grasp the vital principle motivating Christian history, which was not exhausted at its origin. "The gospel," he claimed, is "a living faith, concrete and complex, whose evolution proceeds without doubt from the internal force which has made it enduring, but none the less has been, in everything and from the beginning, influenced by the surroundings wherein the faith was born and has since developed."[20] The claim entailed Loisy's further belief that the historical Jesus was distinct from the stories of Christ. The actual Jesus was not entirely aware of his teachings, which came to find meaning after his time on earth. The explanatory purchase of Loisy's method was to reconcile contradictory elements of scripture, particularly chronological inconsistencies found, for example, in the Gospel of John, which recounts Jesus performing miracles in Jerusalem during different Passovers.[21] Loisy argued that historical distance was required to interpret Jesus's message properly: "The glory of the risen Lord threw new light on the memories of His earthly career."[22] The significance of the Gospels, moreover, depended on their adaptation to the exigencies that the church faced in different historical periods. The very lifeblood of the church was therefore subject to mutation.

Loisy was not the first to tackle church history. Within Catholicism, John Henry Newman (1801–90) of England recognized the discrepancies between the Vatican's authority in the nineteenth century and the birth of Christianity in antiquity. According to Newman, the Roman Church had not corrupted its sources—as Protestants often claimed—but instead added detail to teachings already latent in the fathers' earliest statements. The development of doctrine, as he called it in an 1845 treatise, explained the continuity of sacred scripture over time.[23] Outside Catholicism, Ernest Renan advanced the blasphemous claim that Jesus and Christ were distinct in his *Life of Jesus* (*Vie de Jésus*) (1863). Renan argued that Jesus, born a Jew, cleansed himself of Judaic traces in becoming Christian. Loisy went further than both in his evolutionary approach to church history, which took Catholicism to be shot through with discontinuities.

Loisy ignited a firestorm of controversy in Catholic France because of the prominent positions from which he leveled his heresy. Loisy began his scholarly career at the newly opened Institut Catholique de Paris, before being dismissed in 1893 after teaching seminars on the historical lives of saints.[24] Loisy took a subsequent position at the École des hautes études, where he synthesized his historical method, and garnered the Vatican's disdain for having catalyzed Catholic modernism.

The modernist crisis in the Catholic Church reached a fever pitch in 1907 when the Vatican condemned the movement in a series of formal statements. On April 17, Pope Pius X released an allocution vilifying modernism. Soon after, on July 3, he signed off on the Holy Office's encyclical "Lamentabili Sane Exitu" ("With Truly Lamentable Results"). It featured a list of sixty-five heresies of modernism. Loisy was a clear though unacknowledged target. Instead of reading the Bible as a series of factual propositions, modernists, according to "Lamentabili," treated scripture as an archive of human authors' beliefs formulated at the time of its writing. The Vatican honed in on the claim that "the exegete must first put aside all preconceived opinions about the supernatural origin of Sacred Scripture and interpret the [the Bible] in the same way as another merely human document."[25] In addition to condemning historicist methods of biblical interpretation, the Vatican also shined a light on the modernists' sanguine reception of psychology and neurology. Pius X condemned the belief that "since the deposit of Faith contains only revealed truths, the Church has no right to pass judgment on the human sciences."[26] Only the subservience of science to religion would secure the authority of the latter.

The Vatican's condemnation of modernism came amid the church's waning authority. In 1905, French parliament passed the law separating church and state, which the Vatican received as the final blow of the Third Republic's anticlericalism. The government's decision not to support religious cults economically or socially stoked the flames of the controversy.

On September 8, 1907, Pius X brought the full weight of the Vatican's authority down on "these very Modernists who pose as Doctors of the Church, who puff out their cheeks when they speak of modern philosophy," in the encyclical "Pascendi Dominici Gregis" ("Feeding the Lord's Flock").[27] The modernists' central commitment, "Pascendi" stipulated, was to the doctrine of evolution, or what Pius X characterized as the principle of "vital immanence": "that religious formulas, to be really religious and not merely theological speculations, ought to be living and to live the life of the religious sentiment."[28] According to this view, religion is above all an experience confined to (and therefore immanent in) our subjectivity. The immanent view of church history distinguished *real* history, which follows the chronological sequence of human actions, and *internal* history, which inheres in the pious contemplation of believers. Whereas the former unfolds gradually over time, the latter reflects a transhistorical momentum adapted to the cultural demands of particular eras. The division between two strata of history hewed to spiritualists' division between two dimensions of selfhood: intellection and lived experience (or what Loisy called "sentiment"). What made immanence *vital*, in the eyes of the Vatican, was the idea that faith exceeds believers' intellectual knowledge of doctrine; faith surges from the dynamic interiority of the person and bursts forth in the embodied modalities of devotional practices. The Vatican denounced the critique of intellectualism, instead defending its fortified neo-scholasticism. "What does this *experience* add to sentiment?" Pius X asked. "Absolutely nothing beyond a certain intensity and a proportionate deepening of the conviction of the reality of the object. But these two will never make sentiment into anything but sentiment, nor deprive it of its characteristic which is to cause deception when the intelligence is not there to guide it; on the contrary, they but confirm and aggravate this characteristic, for the more intense sentiment is the more it is sentimental."[29] The condemnation identified what modernists themselves had not. "The consequences of *Pascendi* were devastating," Roger D. Haight affirms; "at once there were no single 'Modernists' who recognized their positions integrally represented in that document, and yet the 'Modernists' were everywhere."[30] "Modernism" was thus a negative identity, a pejorative that the church ascribed to certain thinkers, rather than a theological commitment embraced by its advocates.

"Pascendi" concluded with a series of concrete initiatives designed to weed out modernist influences in the church, including expulsion from seminaries and universities. Consequently, Loisy was excommunicated March 7, 1908. "In the excommunication which thus set me free," Loisy wrote in his autobiography, "I found but a single defect; it arrived twenty years too late!"[31] He was not alone. The Vatican censored troves of modernists. Pius X buttressed his encyclical three years later with the antimodernist oath, "Motu Proprio—Sacrorum Antistitum."[32] From its debut on September 1, 1910, the oath enjoined all clergy to affirm the neo-scholastic doctrines and to declare that God's existence was demonstrable by the light of natural reason.

It was in this context that the critique of intellectualism became socially pertinent. The modernist crisis set the terms in which reason and action, selfhood and science, were contested across French Catholic circles. Those contestations endured for over half a century. Their resolution came about only in 1962 when the Second Vatican Council met and Pope Paul VI subsequently rescinded the antimodernist oath. The council embraced ecumenism, dialogue, and a return to biblical primary sources for Catholic schools. Although the church never championed modernism, the crisis breached at the turn of the century opened the gates for French spiritualism to influence the fate of Catholicism in Europe.

Blondel, Le Roy, and Pragmatist Theology

Blondel and Le Roy entered the École normale supérieure at a time when the collision between Catholicism and the human sciences was all engrossing. Blondel had come from a bourgeois Catholic family in the town of Dijon. He initially entered the Université de Dijon in 1878 to pursue a degree in science and law, but, driven by his true passion, he transferred to the École normale to study philosophy in 1881.[33] Blondel found the institution stifling. The faculty was caught between a public commitment to intellectual exploration and suppressed Christian convictions. Le Roy arrived in 1892. He had been raised in a sailing family that moved to the capital from the coastal town of Le Havre. As I discussed in the previous chapter, Le Roy's studies were in mathematics; he earned a doctorate in the subject in 1898. But just down the hill at the Collège de France, the young student attended Bergson's public lectures. Bergson's thought provided a framework, Le Roy believed, for bringing his Catholic faith into a renewed connection with psychology. Both Le Roy and Blondel did so in an academic milieu where professors guarded their religious beliefs.[34] Nonetheless, they brought pragmatism and theology into dialogue with incredible philosophical gusto, injecting a new phase of spiritualist thought into the tumultuous disputes over the fate of Catholicism in France.

At the École normale, Blondel studied under Émile Boutroux, himself a devoted yet circumspect Catholic. Boutroux's faith left a legible mark on his critique of mechanistic science in *The Contingency of the Laws of Nature*, intent as he was to leave open the possibility of miracles in the causal order of the material world.[35] Although he did not take up God as an explicit theme in his work, Boutroux was aware that Blondel sought to do so in the face of the anticlerical politics of the Third Republic.[36] Boutroux counseled the young Blondel, as he was preparing to sit the agrégation in philosophy: "I still believe that to please the jury, you needn't sacrifice any of your convictions; you can and should remain yourself."[37]

Both Le Roy and Blondel approached religion through the prism of practice. Blondel did so in his doctoral thesis, *Action* (*L'action*) (1893). His aim was to draw on psychology in order to expand the bounds of religious experience. The subtitle of the dissertation, *Essay on a Critique of Life and a Science of Practice* (*Essai d'une critique de la vie et d'une science de la pratique*), gestured toward his argument that selfhood finds its fullest expression in practical engagements with the world. Hence the problem he broached on the first page, "Yes or no, does human life make sense and does man have a destiny?"[38] The problem set in motion a dialectic between the requirements of knowledge and the impulsion of volition. At first blush, action confounds my thought, he argued; one finds it impossible to explain the necessity of acting. "If I try to evade decisive initiatives, I am enslaved for not having acted. If I go ahead, I am subjugated to what I have done." What appears to be an impasse for rational thinking dissolves before the dynamism of our embodied activities. "In practice, no one eludes the problem of practice; and not only does each one raise it, but each, in his own way, inevitably resolves it."[39] Along the similar lines, Le Roy articulated a variant of pragmatism that was part and parcel of the "spiritualist positivism" that I explored in the chapter five. "In short, the new philosophy is neither . . . a philosophy of emotions nor a philosophy of the will: it would instead be a philosophy of action. Now action," Le Roy clarified, "without a doubt implies emotions and the will: but it also implies something else, in particular, reason. . . . *The light and truth should be, in the end, sought after in the very movement by which action unifies the soul.*"[40] The stakes motivating both thinkers followed from spiritualists' engagements with psychology and neurology more broadly. Blondel and Le Roy aligned their accounts of pragmatism with concepts of habit, exertion, and duration, which underscored the spiritual dimensions of embodied volition. What's more, they articulated the creativity of action by mobilizing the very methods that cast doubt on the possibility of free will. As Blondel forcefully argued, "to show that what is transcendent and strange to . . . the positive sciences is the very thing that makes them possible and applicable will be to bring to light what, in science itself, requires that science be surpassed."[41] The conceptual surplus within the human sciences left a metaphysical remainder animating both thinkers' accounts of action.

The choice of action for Blondel's dissertation topic reflected the influence of his teacher Léon Ollé-Laprune (1839–98). The Catholic philosopher at the École normale wrote *On Moral Certainty* (*De la certitude morale*) (1880), in which he argued that matters of faith depend not just on true understanding but also on the inflection of the will. For Ollé-Laprune, religious devotion manifested most clearly in volition (although he stressed that the faithful bring their thoughts and actions into harmony). In drafting his dissertation, Blondel elaborated on what was so ethically vexing about action: I go on acting, he said, despite my inability to answer the problem, why must I act? The problem traversed the entirety of *Action*. Blondel demonstrated that the two horns of the problem, thinking and willing, find their seamless harmony in the theological. Such is especially the case for the human activity involved in scientific practice. Blondel borrowed from his teacher the crucial insight: "To believe is not to make an affirmation simply on the strength of extrinsic reasons, or to attribute to the will the arbitrary power to go beyond understanding."[42] Belief depended on our protean engagements with worldly affairs; "it is . . . to treat Truth as Living or even as a Person . . . to understand that this Living Truth is not only an object of inquiring science or belief, but that it demands trust and the mutual gift of self."[43] For Blondel, the same was true of scientific practice. Its success hinges on a belief in scientific methods. The scientist must invest his or herself in inquiry as a pragmatic—and not exclusively rational—endeavor.

Le Roy also tacked the problem of action, which he posited as the foundation of metaphysics as well as science. Science, he argued, refines our pragmatic engagements with a heightened degree of abstraction and analysis. But metaphysics breaks with the body's natural action-oriented disposition. Le Roy thus identified the cleavage in human experience where the creative activity of philosophizing leaps beyond the practical orientation of science. According to this genetic method, the life of spirit develops out of yet surpasses nature. "Extracting an analysis of the fundamental postulates of common or scientific discourse, cutting off from any point of view that is not that of pure speculation, seizing the internal soul of things in its concrete richness and living originality, and bringing about the supreme unity of knowledge and life: these are [philosophy's] ultimate mission."[44] Le Roy's pragmatism relied on the premise that philosophizing is not a pragmatic activity; it is a creative activity that restores the continuity of experience fractured by the sciences' analytic categories.

Blondel had committed his early writings to science. His first article appeared in the *Philosophical Review*.[45] It was a study of the optical difficulties experienced when perceiving stars on the horizon. He observed that stars appear larger on the skyline than they do when perceived directly above. Blondel leaned on Boutroux's theory of contingency, arguing that astronomy was inflected by the errors of human observation as well as the natural instability of celestial illumination.[46] When it came to his dissertation, *Action*, he brought science into conversation with theology. Whereas Le Roy took the fecundity of embodied practice to be the origin of scientific analysis, for Blondel it was the aporias within science that exposed human action to divine intervention. Blondel argued that scientific knowledge in its broadest sense, from the *sciences dures* to the *sciences humaines*, depended on revelation. "'Christian philosophy' does not exist any more than Christian physics. . . . Philosophy, that is to say, is applicable to Christianity in so far as Christianity exercises, in the last analysis, control and judgment even over men who are ignorant of it or reject it."[47] Unlike religious thinkers who safeguarded the theological beyond the bounds of the scientific by cordoning the explanatory limits of each, Blondel took philosophy to cast light on the foundational problems of science so as to open the possibility of their resolution in theology. Theology was not limited to the otherworldly. For Blondel, it suffused our quotidian life in this world.

Blondel's method in *Action* was threefold. He proceeded, first, by laying out theoretical discontinuities within the sciences; second, by revealing how each finds reconciliation in action. As he quipped, "The positive sciences are not sufficient for us, because they are not self-sufficient."[48] In the third phase, action ultimately depends on our encounter with God.

Blondel's affinity with French spiritualism stood out in his discussion of the sciences' intrinsic discontinuities. The first discontinuity held between what he called the "qualitative" and "representative" dimensions of sensory experience. The qualitative dimension belongs to a researcher's personal interactions with material phenomena—that is, the immediate experience one has in the laboratory. The representative corresponds to the rational character of empirical data; they are mediated by symbols. The two dimensions, Blondel insisted, could not be isolated from each other. They are fundamentally intertwined as two aspects of the same scientific labor. As Michael Conway argues, Blondel drew a phenomenological, rather than ontological, division between the qualitative and representative dimensions.[49] In other words, they accord with composite aspects of scientific practice. Blondel thus disentangled what was often taken to be unitary in the sciences.

The second (and principal) discontinuity held between the mathematical (or exact) sciences and the experimental sciences. The former were inaugurated in the Pythagorean geometry of antiquity. They depend on the mind's autonomous constructions, the ideal realm of reason. The experimental sciences, developed in the Enlightenment with Descartes's and Newton's laws of motion, depend on the observation of nature. Blondel cut the methodological distinction along the lines of analysis and synthesis. Whereas the mathematical sciences analyze a whole in terms of its quantitative parts, the experimental sciences synthesize wholes by selecting among their elements. On the one hand, for example, one could deduce the interior degrees of a triangle (180°) from the three angles subtending a circle. On the other, chemical substances could be identified by means of analogy and verisimilitude among diverse compositions of matter. This was the method of the Russian chemist Dmitri Mendeleev (1834–1907), whose standardization of the periodic table of the elements in 1869 was one of the most significant adaptations of mathematic precision to the study of nature. Boutroux recognized the importance of the discontinuity: "These analyses are incredibly thorough and very penetrating," he wrote in a review of Blondel's manuscript; "certain questions, such as those addressing the relations between the mathematical and experimental sciences, are dealt with in a truly new and instructive fashion."[50] Blondel demonstrated that the mathematical and experimental sciences corresponded to distinct orders of scientific methodology, which mutually borrow from each other in scientific practice. "In spite of their ideal and detached character, the exact sciences have no reason for being and no possibility of existence unless, from the beginning, they implicitly tend to become what they are more and more, a substitute for experimental knowledge and an auxiliary of practical activity."[51] His point was that the precise connections between the methods were aporetic, irresolvable solely in theoretical terms.

As Blondel saw it, such cross-pollination pervaded the sciences. Strikingly, he took evolutionary biology to be emblematic. Blondel argued that Darwin had borrowed mathematical models in order to establish the lines of descent from existent species to their antecedents. Blondel likened the biologist to an alchemist, who strives to transform ordinary materials into a greater composite (i.e., gold). The biologist assumes that species divergence has transitioned gradually toward its present iteration. He or she has no right to do so on the basis of experimental methods alone since the biologist never observes new species' genesis. In place of observation, the biologist implicitly poaches a principle of continuity from mathematics, namely that of a mechanistic continuum. Species mutations are thus arranged in linear fashion, moving by divergent transformations from the past to the present. Blondel's critique of evolutionism targeted its underlying mathematical epistemology, a framework that eluded biological explanation.

Although the discontinuities between mathematical and experimental science took root in the mind of the scientist, neither could be mended by mental operations alone. "Our power always goes further than our science, because our science, risen from our power, needs that power still to find in its support and its end."[52] According to Blondel, it was up to action to settle methodological discrepancies. He drew the consequence that scientific methods constitute conventions. For Adam C. English, Blondel advanced a narrative account of science. "His appeal to 'the science of action' was not to establish a new foundationalism but to undercut late nineteenth-century scientific foundationalism by acknowledging that all sciences presuppose certain fictions."[53] Scientific knowledge was subservient to scientific practice since the will creates the conventions of scientific communities. As Blondel claimed, "The positive sciences are only the partial and subalternate expression of an activity that envelops, sustains, and overflows them."[54] Far from having condemned scientific methods for being flimsy constructions, Blondel endowed science with pragmatic support and authority. In our twenty-first century, decades of social constructionism in social theory and science studies have rehashed the premise; but it is worth keeping in mind just how revelatory it was in 1893 for Blondel to undermine the mythical imperative that science discover an essence of nature. "Thus are removed the shackles that once subjected the sciences to the fictitious necessity of faithfully representing and entirely constructing an objective world, a world they were able to present to minds charmed by the certitude and the precision of their results as if it were reality itself, a world of prestigious mechanisms where one could not tell whether it was the work of the senses or of reason."[55]

Similarly, Le Roy argued that units of measurement (e.g., weight, dosage, speed, or purity) constitute conventions. He emphasized that they flow from our body's innate action-oriented disposition. "Positive science can thus be defined: an immediate prolongation of common sense."[56] The claim was of a piece with Henri Poincaré's argument that scientific concepts depend on intuition in the contexts of discovery as well as justification. Our working categories are only refined—but not created—by the understanding; they must be adapted to particular problems. "What [mathematics] have gained in rigor, they have lost in objectivity. It is by distancing themselves from reality that they acquired this perfect purity."[57] In his mathematical work, Le Roy lent support to his teacher, showing that the symbols of Euclidian space are not necessary postulates of reason. Any number of alternative symbols, including non-Euclidean concepts of space, could also meet the pliable demands of scientific reasoning. For Le Roy as for Blondel, scientific concepts were fit to our pragmatic engagements with the material world.

The third discontinuity that Blondel diagnosed in *Action* held between the methods and objects of experimental psychology. In accord with Le Roy and fellow spiritualists, he believed that an external approach to the mind immobilized its activity and thus failed to adequately capture the internal initiative of psychological activity. "In the very use of his 'subjective method' [the psychologist] considers the subjective as fact and not as act; he disfigures it under pretext of studying it."[58] To overcome the discontinuity, Blondel argued that psychology must adopt "not the static viewpoint of the understanding, but the dynamic viewpoint of the will."[59] Following his phenomenological method, Blondel urged psychologists to explain the activity of consciousness from within: "A science of the subjective will inevitably be a mental dynamics."[60]

His vision of a comprehensive psychology was one that took the life of spirit to be complementary with clinical and experimental inquiry. Together, they would explain the dual dimensions of action. On the one hand, the will injects "infinity" (read: contingency) within the finite world of nature: "The infinite, that means here what surpasses every distinct representation and every determinate motive, what is without common measure with the object of knowledge and the stimulants of spontaneity."[61] On the other hand, one experiences his or her own willing as a chasm stretching between the capacity of action and particular acts—what Blondel respectively called the "willing will" and the "willed will." My capacity for volition inevitably appears greater than *what* I choose to do. For instance, it is rare that one's artistic creation leaves a lasting impression on others' minds, achieving an enduring standard of beauty. Equally fleeting is the construction of a scientific law, which achieves explanatory repetition across diverse contexts over time.[62] Our desire for such accomplishments strives for "as perfect an agreement as possible between the *élan* and the result of our effort, equality between the amplitude of the voluntary aspirations and the magnitude of the willed ends."[63] We find in Blondel's treatment of the will his turn away from the dominant Thomistic philosophy of the period and a return to Saint Augustine. In the first book of the *Confessions,* Augustine stated that we feel ourselves to live in the image of God because our hearts are distraught by a restlessness to return to Him.[64] Only heaven would provide relief. Herein lay the religious dilemma at the heart of action: the restless heart of a continuously deferred urge to harmonize my infinite power of becoming with the finite results that I bring into being. Pragmatic psychology thus spilled onto a theology of the will.

Blondel arrived at the fundamental tension of the will after analyzing the discontinuities of the natural and human sciences. The division he discerned between the infinite power of volition (freedom) and the finite results of its actions (necessity) overcame the intellectual picture of selfhood in two senses. To begin, Blondel did not predicate the problem of action on the external duality between the effects of one's actions in the world and the private intentions within one's subjectivity. He eschewed such a view of ethics, which would seek to bring one's objective output under subjective mastery. There is instead an inner heterogeneity at the heart of action. It is this infinity within our finite will that lays bare the divine grace at the core of our personhood. Second, Blondel marshaled the inner heterogeneity of action to upend Kantian epistemology. Kant had argued that reason steers off the rails, as it were, when it deviates from the straight track of rational scientific inquiry. Whereas the first *Critique* diagnosed the contradictions—what Kant called "antinomies"—resulting from speculation beyond the limits of reason, Blondel illuminated three contradictions *within* scientific reasoning. "All the antinomies some have claimed to find in the realm of speculation reside in science itself," Blondel averred. "Kant makes metaphysics depend on . . . the possible agreement between the *a priori* of the analytical syntheses and the *a posteriori* of the synthetic analyses. But it is in the sphere of the sciences that the duel and the reconciliation takes place, without our having to consider phenomena otherwise than as phenomena or to suppose anything else beside or underneath them."[65] In the eyes of Blondel, Kant had construed metaphysics as fanciful and science as its antidote because he was stuck in the grips of an intellectual picture of the self.[66] In response to rational epistemologies of both ethics and science, Blondel demonstrated the antinomy at the core of selfhood: the necessity of freedom.

Here Blondel surpassed Le Roy and fellow spiritualists in his appeal to the divine as the ground of scientific methodology as well as philosophical speculation. He argued that the human and life sciences did not only generate a conceptual surplus—a surplus that would be the fodder for metaphysics, and in turn, theological revelation. In Blondel's eyes, theology was required by science. God is the energetic wellspring from which human action draws its power. The infinite that we feel in the willing will is His. "Wherever we stop, He is not; wherever we walk forward, He is. It is a necessity always to go further, because He is always beyond our action."[67] Although he invoked a spiritualist characterization of volition, Blondel went further by naming God as the creative power that animates human freedom: "It is because action is a synthesis of man with God that it is in perpetual becoming, as if stirred by the inspiration of an infinite growth."[68] Our finite will never amounts to the antecedent initiative of God's infinite will. It is because of God that action confronts a chasm between the willing will and the willed will. In this respect, Blondel's formulation echoed Kierkegaard's maxim, "If there were nothing eternal in man, he could not despair at all."[69]

This chasm at the heart of action lay bare what Blondel called the "one thing necessary." The term signified that which we cannot fully explain: the necessity of our freedom. We confront the "one thing necessary" not in the form of a positive resolution of the problem, but instead as both a constraint on and a choice of our action. Understood as a constraint, it manifests as the weight born by the will: the insurmountable demand that action escape its finitude. "The 'one thing necessary,' then, is not the obscure side of my thought, the invisible reverse side of my consciousness and my action, as if I were to see it only within myself and as if all its reality consisted only in the idea I have of it."[70] According to Blondel, God does not retreat from the world. He permeates our practical life. For the will ineluctably tends toward the transcendent. As Blondel claimed, man aspires to be God. But it is a self-surpassing aspiration. This philosophical constraint gives way to a practical choice of our action: "to be God without God and against God, to be God through God and with God, that is the dilemma."[71] Understood as a choice, the "one thing necessary" poses a practical question of us, a question posed not in the form of a "yes" or "no" interface but as an animating force taken up by my action. On the one hand, I can restrict my will to the material order: that I am the author of my own life. In so doing, I make my own action the point of departure and arrival for my destiny. In this case, I deny God. I can spurn the weight my will bears; but I fail to reconcile the tension between my inescapable demand for the infinite and my feigned conceit for the finite. On the other hand, I can acknowledge the finitude of my action and, ultimately, my dependence on God's infinite power. "Reduced to its own resources alone, [the will] can only acknowledge its ignorance, its weakness and its desire, for it is true to its infinite ambition only inasmuch as it recognizes its infinite powerlessness."[72] Blondel's pragmatism therefore paved the road to necessity, a divine necessity that finds reconciliation with freedom in accepting the gift of God's grace.

Had Blondel reduced faith to a psychological phenomenon? Were readers of *Action* left to understand Catholicism as merely a personal exercise? This was the argument of Abbé Charles Denis (1860–1905), who rebuked Blondel for having aligned his thought with the modernist accommodation of religion to secular practice. Denis was the head editor of the *Annales de philosophie chrétienne*, the chief organ of academic theology in France. From his pulpit, Denis targeted *Action*, claiming the book's purpose "is to put Christian apologetics on a psychological terrain."[73] Apologetics is a form of essay written to support faith on a rational basis and defend it against misrepresentations. But this was only partially Blondel's aim. As he elaborated in his *Letter on Apologetics* (*Lettre sur l'apologétique*) (1896), his wider objective was to argue that a philosophical intervention into religion had to respond to the exigencies of modern philosophy, which were imbricated with the natural and human sciences. Hence Blondel's full title, *Letter on the Exigencies of Contemporary Thought in Matters of Apologetics and the Method of Philosophy in the Study of the Religious Problem* (*Lettre sur les exigences de la pensée contemporaine en matière d'apologétique et sur la méthode philosophique dans l'étude du problème religieux*). Blondel showed that philosophy conducts a natural analysis of consciousness, whereas theology tends to the supernatural domain of religion. The distinction, he enjoined, ought not be taken for granted. Blondel mobilized the conceptual tools of philosophy in order to clear the ground for theology, and reciprocally, to bring the problems of religion into the realm of philosophy. Doing so called for a new epistemological framework, one that would upset what he called the notion of "immanence," the idea "that nothing can enter into a man's mind which does not come out of him and correspond in some way to a need for development."[74] As long as philosophy, in the wake of Kant's *Critiques*, remained a hermetically sealed inquiry untouched by the supernatural, it would fail to account for the external constraints on our thought. "For no science can be precisely defined unless we recognize the presence and, as it were, the pressure, of a limit beyond which other perspectives lie open, determining both what it is and what it is not."[75] Blondel argued that philosophy and theology, instead of constituting absolutely separate domains, meet in what he called "immanent transcendence." Philosophy illuminated the "the blank spaces which cannot be filled in or established in their reality by any resources of ours" and thus offered theology its point of departure.[76] Its task was neither to formulate a concept of God nor to justify His existence. "[Philosophy] cannot therefore pronounce on the question of fact," Blondel argued; "it can only determine the dispositions which prepare for the understanding of facts and for the practical discovery of truths which emanate from another source."[77] A philosophical psychology of action would point toward the pragmatics of faith.

The pragmatism of Blondel's and Le Roy's early thought developed out of their overlapping engagements with science, philosophy, and theology. Where Le Roy invoked pragmatism to explain the embodied bases of philosophy and science, Blondel advanced a pragmatism that transcended both. Their differences, I want to suggest, shine a light not only on the Catholic inspirations of spiritualist thought, but also on its theological limitations. Readers certainly found in Le Roy's work a conceptual assessment of religious motifs. It was not, however, theology. By way of contrast, audiences found in Blondel's work both fruitful dialogical partners who belonged to the Catholic tradition as well as a resolutely theological ambition: to steer the sciences into contact with religion. As Jean Leclercq argues, Blondel articulated a new ontological proof of God.[78] The Western Christian tradition since Saint Anselm had deduced God's existence from the definition of God: if we have a coherent belief of a perfect being, then existing must be one of its attributes. Blondel established the presence of the divine not on the basis of abstract thought, but on scientific practice. "Instead of looking for the necessary outside the contingent itself, as an ulterior term, it manifests it within the contingent itself, as a reality already present."[79] For Blondel, Christian spirituality imbued the material world. For Le Roy, spiritualist concepts inflected Christianity. These thinkers' divergence sharpened when both turned to the Catholic problem of dogma.

The Problem of Dogma

Confronting a rising tide of secularism, the Vatican adapted its position in France. In the 1890s, Pope Leo XIII initiated the *ralliement*, an official campaign that reconciled the church to the Third Republic. French Catholics had allied with conservatives in anxious defense of the social order, which they thought had been disrupted by Republicans' anticlerical laws. Divorce was legalized (albeit within limited parameters); the army could no longer participate in religious processions. Overall, attendance at mass as well as the taking of Easter communion declined.[80] In response, the Vatican preached tolerance. Leo XIII penned his encyclical "On the Church and State in France" ("Au milieu des sollicitudes") (1892), in which he enjoined

Catholics to work with the Republic, if not for its politics then at least for its prosperity.

Reconciliation was tendentious, especially amid the national drama of the Dreyfus Affair. In 1894, Alfred Dreyfus, a Jewish artillery captain, was accused of treason for having transmitted sensitive information to the Germans. An army court reached a guilty verdict. Deemed a traitor after a memorandum purported to be his was found in a trash bin, Dreyfus was exiled on Devil's Island (just off the coast of French Guiana). Four years later, the case burst into the public limelight. It was made known that Dreyfus had been framed. Throngs of sympathetic supporters painted him as the victim of an anti-Semitic cover-up. The spectacle of the affair tore apart the "two Frances." Catholics largely sided with the military (while equally suspicious of Jewish conspiracies in the government); Republicans rallied on behalf of Dreyfus (while skeptical of Catholics' influence over politics). Priests' and bishops' anger at the Third Republic reignited. Their situation worsened when a new government of radical Republicans came to power in 1899 under the prime minister Pierre Waldeck-Rousseau (1846–1904). The Associations Act of 1901 followed, a law that eliminated restrictions on organizations with the exception of churches, whose property was subject to public administration. Freedom of association did not wholly apply to religious association. In 1902, the radical Émile Combes received a wide electoral mandate to take over the government and to enforce the Associations Act strictly. Anticlerical policy continued to mount, culminating in 1905 with the official separation of church and state.

When Pius X assumed the papacy in 1903, he refused to appease the French government. He condemned not only Republicans' policies but also what he took to be their philosophical motivations. In his list of sixty-five of heresies of modernism, Pius X specifically condemned two principles: pragmatism and history. In the first case, the Vatican targeted those who believe that "the dogmas of the Faith are to be held only according to their practical sense [*regula praeceptiva actionis*]; that is to say, as preceptive norms of conduct and not as norms of believing."[81] Blondel and Le Roy were among the unacknowledged targets. In the second case, Pius X took issue with the historicist method of biblical exegesis: "Like human society, Christian society is subject to a perpetual evolution."[82] From the perspective of Rome, modernists embraced unsteady conceptions of selfhood, which ceased to hinge on the certainty of reason and the proofs of revelation but hinged instead on practical participation in a fluid social world.

The problem of dogma and history lay at the heart of the modernist controversy in France. Dogmas had been declared since the biblical era of the Apostles, and they continued to be significant for the modern church. The dogma of transubstantiation, for example, was at the center of the rift between Lutheranism and Catholicism during the sixteenth-century Reformation. While both sects recognized the real presence of Christ in the Eucharist, Catholics believed that communion transforms bread and wine into Christ's body and blood, whereas Luther held transubstantiation to be symbolic. In the late nineteenth century, Pope Pius IX defined the Immaculate Conception of Mary, her birth untainted by original sin, as a dogma.[83] And as recently as 1950, Pope Pius XII promulgated the additional dogma of Mary's Assumption, her bodily and spiritual ascendance to heaven upon death.[84] The problem of dogma in early twentieth-century France threw into question the very possibility of revealed truth. On which grounds could dogmas be held as true in the face of modern science? What kind of data could dogmas be said to constitute? And how could Catholics continue to defend their beliefs given the ascendant scientific demands of empirical verification and experimental falsification?

The questions of dogmas and historicism were intertwined. Since the church had evolved since the time of Jesus, modernists believed that the dogmas he proclaimed did not remain the same over the course of two millennia. "He demanded faith in the approaching kingdom," Loisy wrote, "but the idea of the kingdom and of its proximity were two very simple symbols of very complex matters, and even those who were the first to believe must have attached their minds more to the spirit than to the letter."[85] The separation between the history of the Gospels and the evolution of the church left the meaning of dogmas open to debate. Do dogmas encapsulate the truth of miracles? Or might they be mere symbols whose expressive force transforms over time? Loisy's answer was unequivocal: "The Church does not exact belief in its formulas as the adequate expression of absolute truth, but presents them as the least imperfect expression that is morally possible; she demands that man respect them for their quality, seek the faith in them, and use them to transmit it."[86] Loisy dared to claim that the historical development of the church confirmed the pragmatic, rather than absolute, value of dogmas. They amounted to allegories.

In their effort to reconcile dogmas with the authority of science, Blondel and Le Roy delved deeper into the modernist controversy. Blondel first tackled the problem in 1904 with a series of articles, "History and Dogma" ("Histoire et dogme").[87] At stake was the problem of how to relate the facts of Christian history to beliefs that were central to the church's teaching without reducing one to the other. How, that is, did the Bible ensure the church's authority at the same time that the church interpreted and supported the Bible?

Blondel sought a rapprochement. His aim was to guarantee at once that dogmas constitute more than faint echoes of the Gospels' original testimony and that the church participates in and inflects a sacred history more expansive than the sum total of its teachings. The challenge, according to René Marlé, amounted to articulating an interpretive principle that "is at the same time that of the relations of Scripture to the Church, and that of an eternal truth given 'once and for all' the manifestation of which develops in time."[88] Blondel called this principle "tradition." He claimed that tradition is a "living synthesis" of the collectivity and the individual, embracing "within it the facts of history, the effort of reason, and the accumulated experiences of the faithful."[89] The value and purpose of dogmas, Blondel averred, is to convey tradition.

Blondel did not defend a fossilized notion of tradition. In fact, he was motivated to overcome what he characterized in an earlier essay as Thomists' "static description of the rational and proprietary elements of dogma."[90] His articles on dogma and history elaborated on these schematic remarks. "So far is 'development' from being heterodox," he wrote, "that it is the static idea of tradition, fixism, which is the virtual heresy."[91] Blondel coupled his idiosyncratic account of the church's history with a critique of the thesis of historicism, and of Loisy's version in particular. Blondel argued that modernists saw history as a positivist accumulation of facts; further, the principle of evolution, which Loisy conceived as Christians' "living faith,"[92] reflected only the external pressures that the church confronted over time. Blondel crafted his account of tradition to mend the division that historicism tore open between the miracles preceding the founding of Christianity and the church's subsequent development. "If one accepts [historicism]," Blondel wrote, "although one might continue to say that Christianity is founded *on* Christ as a cathedral is built on a geological foundation, one can no longer add that it was explicitly founded *by* Christ, because one no longer looks behind the historical facts for a substantial and active reality."[93] As long as history remained confined to facts alone, then the history of Christianity, Blondel argued, would constitute one chronology among others. For Blondel, historicism furnished a wholly inadequate justification for the authority of the church's teaching.

What, then, is the status of the church's history? And on which basis is it subject to modification without severing the thread connecting Jesus's ministry to contemporary Christians' belief in salvation? Blondel invoked his philosophy of action to resolve these problems. Tradition reflects the infinite action of Christian history, of which the scriptures are only finite expressions: "[Tradition] relies, no doubt, on texts, but at the same time it relies primarily on something else, on an experience always in act which enables it to remain in some respects master of the texts instead of being strictly subservient to them."[94] Blondel claimed that the collective action of Christianity surpasses the particular acts crystallized in dogmas. In a letter to Blondel, Boutroux recognized that his former student elevated his thesis in *Action* to the animating principle of Christian history: "It is indeed from within that the facts must be acted upon. It is the spirit [*l'esprit*] which is the seat of religion, but it is the spirit of another order than what appears and passes. And its reality, for us, is in fact other than both mind and matter."[95] Blondel thus tied his theological critique of historicism to his philosophical critique of intellectualism in the name of a pragmatist notion of tradition. The formulations of scripture appeal only to the intellectual capacities of believers. Tradition, however, is not transmitted by exclusively intellectual means; it inheres neither in a supernatural world divorced from the natural order, nor in the historian's balance sheet. Tradition is the historical activity that preserves the church's "sovereign authority in regard to Scripture, but in such a way as to leave the Scriptures their own physiognomy and their original spontaneity."[96] It was through his encounter with the problem of dogma that Blondel radicalized his philosophy of action. Dogmas, Blondel argued, manifest the collective and historical action of Christianity. Instead of being only objects of contemplation whose mystery demands rational examination, dogmas are also the living deposits of Catholic history.

Le Roy responded in "What Is a Dogma?" ("Qu'est-ce qu'un dogme?") (1905). His aim was to adapt dogmas to the requirements of modern science so that Catholic faith could be safeguarded from the demands of experimentation and verification. The dilemma, Le Roy surmised, was that Catholics mistakenly opened their faith to scientific criticism by conflating revealed truths with rational truths. Dogmas were too often construed as intellectual propositions, as if they were statements of a theorem, and thus, in the eyes of science, a nonfalsifiable theorem. To resolve the dilemma, Le Roy argued that Catholics ought instead to treat the church's teachings as pragmatic allegories. He thereby claimed to undermine the very ground on which science could contradict Catholic teachings. "The Catholic, obliged to accept them, is not restrained by them except as regards rules of conduct, not as regards any particular conceptions."[97] Freed from the duty of defending a *theory* of God's teachings, Catholics could nonetheless follow His practical rules, unencumbered by scientific reproach.

Le Roy presented his article as a series of interrogations, wary that he might be read as a blasphemous modernist.[98] Le Roy's pragmatist inquiry drew the praise of progressive Catholics such as Loisy.[99] Yet many found the article to be sacrilegious for having evacuated Catholicism of faith (*fiducia*), as when Le Roy claimed, "it cannot be repeated too often that

Christianity is not a system of speculative philosophy but a source and regimen of life, a discipline of moral and religious action, in short the sum total of practical means to obtain salvation."[100] One priest decried Le Roy's "attitude as dangerous to the faith of so many."[101] Another compared Le Roy with such detestable heretics as Gnostics, Manicheans, Waldensians, and Voltairians.[102] Blondel, for his part, suspected that Le Roy had dispensed with theological virtues and recklessly flouted the Vatican. In a letter to Lucien Laberthonnière (1860–1932), the progressive priest and public proponent of *Action*,[103] Blondel confided that he believed Le Roy had gone too far: "I wish to say to Édouard, 'Beware!' You are rolling down a hill that will take you much farther than you think, somewhere I refuse to let myself be taken."[104] Blondel's caution proved prescient. The Vatican promptly listed Le Roy's books on the index, alleging that they condemned belief in the name of practice.

Le Roy did not intend to evacuate the rational content of dogmas entirely. He instead sought to delimit their intellectual and pragmatic dimensions. As part of his critique of intellectualism, Le Roy argued that the rational content derives from dogmas' normative guidance. The dogma of Jesus's resurrection, for example, amounts to a parable: "Be in relation to him as you would have been before his death, as you are with a contemporary."[105] Its intellectual dimension, in turn, serves to guard against heretical notions of Jesus's resurrection. Debates over rational interpretations, in other words, arise amid schisms within the church; it is in these cases that the intellectual dimension of dogmas function negatively to condemn heresies. The clergy and the laity might construe the theoretical meaning of the resurrection differently, "but whether ignorant men or philosophers, men of the first or of the twentieth century, every Catholic has always had and always will have the same practical attitude with regard to Jesus."[106] Although the rational meaning of dogmas participate in the flux of time, their pragmatic instruction, Le Roy argued, endures as a guidepost for the Catholic community.

Blondel believed that Le Roy had pushed his pragmatism both too far and not far enough. Not far enough, because Le Roy conceived the practical value of dogmas in narrowly subjective terms, as if dogmas offer guidance only to individuals; whereas Blondel, to the contrary, conceived dogmas as expressions of collective practice. Too far, Blondel contended, because Le Roy risked liquidating what was distinctly Catholic in church doctrine. He had distorted faith by tethering divine love to humans' autonomous will. Blondel argued that Le Roy neglected the heteronomy by which our individual will depends on God's infinite will: "The more I reflect, the more it seems to me that our divergences are based on this profoundly deep cause: we do not have same idea of the supernatural. It would nearly seem that you only see the supreme blossoming of our own nature, as if it were a crowning achievement of the divine destiny calling us—a destiny which is realized by the harmonious development of our whole being; but this apparent heteronomy turns into a perfect autonomy. For me, to the contrary, the autonomy of our will must accept, even love, the real and irreducible heteronomy of divine love."[107] As Blondel saw it, so long as Le Roy's pragmatic account of dogma remained confined to subjective psychology, then the grace of God's gift became unthinkable. The absolute would be tantamount to our inner psychology. To the contrary, Blondel insisted, divine love arrives as "an intrusion, as a substitution in us of the infinite that expands so far that we cry in pain."[108]

Pragmatism, for Le Roy, was not the same as subjectivism. "Practice," he remarked "does not in the least mean a blind step, without relation to thought or consciousness."[109] He understood religious practice as part of the church history, albeit a history open to change. Le Roy saw himself as expanding the domain of faith beyond the intellect, and thus broadening Catholicism beyond its confines in neo-scholasticism. This was his argument in a 1906 article, "Scholasticism and Modern Philosophy" ("Scolastique et philosophie moderne"), which dismissed the claims of Thomism to speak for the entirety of Catholic doctrine.[110] According to Le Roy, no single doctrine could encapsulate religion's dynamic flux. Explicitly aligning himself with Bergson, Le Roy suggested that Catholicism would never reach a final resolution to the problems it confronts, as Catholic history remains open to the new and unforeseen.

Conspicuously quiet during the debate was the lone non-Christian among spiritualist thinkers. Bergson, born to Polish and English Jewish parents, furnished the conceptual tools taken up by Le Roy. Bergson privately offered Le Roy his approbation for "What Is a Dogma?," which he believed had established a prudent division of labor between the intellectual and pragmatic dimensions of faith. "The article arrives at an important philosophical conclusion, namely, that a dogmatic proposition can be determinate and precise while its meaning carries a share of indetermination. It suffices," Bergson added, "that the proposition relies on a practical attitude, which outlines in a way the *motor articulations* of meaning—the meaning itself remaining (partially) indeterminate."[111] What Bergson saw (and Catholic critics neglected) in Le Roy's pragmatist conception of dogma was its basis in embodied experience. No longer conceptualized as a theory, but instead as a social and biological practice, Le Roy showed at once that faith was compatible with the human sciences yet immune to their atheistic reproach.

Bergson and Le Roy had previously discussed the controversies facing the church at the French Society of Philosophy in 1902. There, Le Roy described Christian teachings as a power, "which presses on reason, but overtakes and judges it, which presses on an intellectual dialectic, but also on what Pascal called the *heart*, that is to say on a progressive experience of life, and which presents what the language of the new philosophy calls the primate of action in consciousness."[112] Le Roy aligned his energetic approach to faith with the temporal movement of experience. In legibly Bergsonian terms, Le Roy suggested that dogmas "are *historical* facts, which is to say facts *capable of duration* and it is in their very duration that we should envisage them to understand them well."[113] Although Bergson did not wade into the modernist controversy, his thought sent ripples throughout.

Indeed, many critics of modernism honed their criticism on Bergson. Writing in the *Revue du clergé français*, the theologian Eugène Lenoble argued that Bergson's spiritualism was not limited to individual psychology. The author recognized that spirit was an impersonal analogue to God, but Bergson replaced His divinity with a vacuous concept, the vital impulse.[114] For similar reasons, the Dominican theologian Réginald Garrigou-Lagrange (1877–1964) named Bergson and Le Roy as the leaders behind the modernist movement.[115] Even Blondel made a point to delineate the "very profound philosophical opposition" between his thought and Bergson's. In a 1905 letter to Laberthonnière, the editor of the *Annales de philosophie chrétienne* (and whose own works had been listed on the Vatican index[116]), Blondel acknowledged his affinity with the critique of intellectualism. Yet, he added, unlike Bergson, for whom "everything is subordinated to utilitarian instinct so as to draw back toward 'intuitive sympathy,' I search in action for the natural prolongation, the application and progressive enrichment of thought."[117] Easier said than done. Many in the church aligned Blondel with modernism. Garrigou-Lagrange, who policed neo-scholastic doctrine from his position at the *Angelicum* in Rome, scrutinized Blondel's writings throughout the remainder of his life.[118]

As the Vatican's condemnations of modernism cast a shadow of paranoia over French Catholic circles, Blondel went into hiding and pseudonymously published his subsequent work.[119] Le Roy, undeterred by the ferocity of his critics, plunged further into the debate in 1907 with *Dogma and Critique* (*Dogme et critique*), an expanded version of the article he had published two years prior. Fresh off the press, Le Roy's treatise ended up on the Vatican index. The fervor surrounding his pragmatic and historicist rendering of doctrine was so vitriolic that Blondel distanced himself entirely from Bergson's disciple.[120]

Bergson's books followed. In 1914, the Vatican condemned Bergson as a modernist whose work upset the neo-scholastic orthodoxy. In his defense, the poet Charles Péguy (1873–1914) published a two-volume essay on Bergson's philosophy, in which he made the case for its Christian inspiration.[121] To Péguy's mind, Bergson offered not so much a theory, but a lifestyle informed by grace, liberty, and the soul's openness to novelty. The true threat to the church was a faith predicated on the false security of mechanized habit. In fact, Bergson made his respect for Catholicism known. In his old age, he requested that a priest be present at his deathbed to give his last rites.[122] Péguy found in Bergson's work the impetus for a revitalized Catholicism that sparked a return to the religion's primary sources. In lieu of the medieval writings of scholasticism, Péguy urged a return to the patristic origins—as he put, "from a less profound to a more profound tradition; a discovery of the most profound resources."[123]

The debates over faith, science, and dogma in the early twentieth century opened up a French variant of pragmatism. At stake was the tenability of religious belief in the face of scientific advancements and social upheaval. At the center, Blondel and Le Roy penetrated the depths of religious consciousness using the very scientific and philosophical methods in which they were originally trained. The result was a more capacious articulation of faith grounded in the totality of human practice.

These thinkers' interventions in the modernist crisis reverberated throughout the twentieth century. In particular, Blondel's work catalyzed *la nouvelle théologie,* which brought about enduring change in the church. By demonstrating how human action is inherently historical, Blondel opened the arc of history to its supernatural destiny. His influence was immense for French Jesuits at the seminary of La Fourvière in Lyon. They found motivation in Leo XIII's call for *ralliement* with the Third Republic and took from Blondel's account of history the lifeblood of a reenergized faith. Notably, Henri de Lubac (1896–1991) made the dynamic between nature and grace into a central preoccupation of Catholic experience in his book *Surnaturel* (1946). From Blondel's writings, Lubac drew the insight that the divine continuously penetrates human action such that knowledge of God is necessary yet never final. For the intellect was shown to be too narrow to master the ebb and flow of tradition. In this light, the limitations of neo-scholasticism appeared even more cramped. In fact, on the occasion of the 1993 centennial of *Action*'s publication, Pope John Paul II recognized that Blondel's "thought and life, was able to affect the coexistence of the most rigorous criticism and the most courageous philosophical research with the most authentic Catholicism, even as he drew on the very fountainhead of dogmatic, patristic, and mystical tradition."[124]

More broadly, the debates in which Blondel and Le Roy took part had a noticeable impact on Catholics' efforts to adapt their faith to the modern conditions of European society. No longer able to participate in political life directly, especially in the aftermath of the 1905 law separating church and state in France, the Vatican left it up to lay Catholics to make religion a meaningful part of private life. What it meant to have a Catholic identity thus came to matter in a profoundly personal way. Action, history, and embodiment became definitive of religious selfhood in the twentieth century. Equally transformative was the teaching of Catholic doctrine. Blondel and Le Roy brought philosophy into dialogue with theology and used philosophy to overcome the rationalism of neo-scholasticism. What ensued was the *ressourcement* movement, which revived primary sources in Catholic education. Péguy's call for a turn away from Thomist manuals and a return to Catholicism' originary moment was eventually a definitive achievement of the Second Vatican Council. When the council closed in 1965, the church embraced changes adapted to the modern world, including religious freedom, human rights, and interfaith dialogue.[125]

For our purposes, the debates among Le Roy, Blondel, and fellow Catholics expose the religious inspirations and limitations of French spiritualism. As much as Catholicism provided an impetus for spiritualist thinkers, it is difficult to find among them a bona fide theology. Indeed, the gaps separating Blondel's and Le Roy's versions of pragmatism offer sufficient testimony. Circumscribed in the logic of the embodied will, Le Roy's account of faith parted with Blondel's much deeper account of divinity. Nevertheless, their shared engagement with pragmatism made possible a rapprochement between Catholic faith and the human and life sciences. Even more, their contributions to a distinctly French pragmatism were possible only within the historical circumstance of the turn of the century. The modernist controversy occasioned these thinkers' intellectual trajectory: from the philosophy of action to the theology of action. Indeed, the problems of historicism and dogma, which shook the church's foothold in France, set the stage for Blondel and Le Roy to articulate in their separate ways a historical and social account of the self's spiritual powers. In so doing, these figures reimagined the dynamics of religious experience at the intersections of science, selfhood, and materiality.

Epilogue

In 1908, Émile Boutroux reflected on four decades that had transpired since the rise of "spiritualist realism or positivism" as it was foretold in Félix Ravaisson's 1867 report on the state of French philosophy.[1] Since then, France had transitioned from an empire to a republic whose scientific institutions and educational system were the envy of Europe. Boutroux remarked that the philosophical establishment, "far from claiming to be self-sufficient, considers that it could only find in the sciences, life, and the arts, as they develop spontaneously, the necessary material for its theories."[2] Although philosophers had come to thrive on scientific literature, their discipline confronted its own self-effacement. "The present movement tends toward the complete abolition of philosophy," he worried, "and its pure and simple replacement by science."[3] As he turned to foretell what was to come, Boutroux was ready to offer a eulogy for the movement that he had once set in motion. But there was a remedy. Reigniting the impetus behind French ideas of spirit could spark an intellectual renewal. By closely following the human and life sciences, and mining the problems within them, philosophers would yield a conceptual surplus: the unresolved remainder being fertile material for thought. Philosophers might thus "find themselves driven to consider, beyond proper facts, those actual realities grasped with the senses, the internal and subjective work of the spirit, the living power that exceeds in reality and richness all the concrete forms through which it appears."[4] If science offered a point of departure, spirit remained a future point of arrival.

The story of French spiritualism reached its climax in the early twentieth century. It was a story of overlapping endeavors to playfully appropriate neurology and psychology, drawing from scientists' tangled lines of inquiry—and their unanswered questions—the resources to craft visions of the self rooted in France's intellectual heritage and open to unforeseen horizons.

That figures from Biran to Boutroux and Bergson championed embodied volition as the liaison between mind and body was hardly a foregone conclusion. Nor was it self-evident that the research agenda of the young brain sciences would provide fecund sources for these figures' philosophical anthropology. Since the seventeenth century, the spiritualist tradition had cultivated a dualist metaphysics. The segregation between spirit and matter made the materialist turn in spiritualism improbable—if not unimaginable. But the spiritual and the material proved to be pliable, productive concepts in the shifting cultural atmosphere of the Third Republic. New journals, scientific laboratories, educational institutions, and professional societies enabled close encounters between professional philosophers and international networks of scientific correspondence. What resulted was a widespread initiative to advance neurology and psychology by redirecting their reductive implications toward an enlarged account of human experience.

Throughout, Kant's legacy was a formidable dialogical partner. Its endurance in France reminds us that spiritualism did not stage a revolt against positivism—nor, to be sure, a rebuke of the ascendant objectivity of scientific methodology. Rather, a wide swath of European thought parted with the parsimonious separation of human freedom from the order of nature. In opposition to the neo-Kantian account of selfhood, spiritualist thinkers refined their pragmatist understanding of the biological and social dimensions of action.

This pivotal juncture in European intellectual history invites us to appreciate the immense sense of awe inspired by science. Driven by neither anxiety nor disenchantment, an integral stand of fin de siècle culture instead sought to transform the norms of scientific validity from within. Neurological and psychological research was to be closely examined and charitably reconstructed. Sensory-motor and quantitative approaches to reflex action, neural association, and cerebral localization became the raw material with which the figures in this book reimagined the nature of knowledge, time, agency, and faith. Indeed, spiritualism fashioned a critical tool kit with which to expand the bounds of scientific inquiry.

I'd like to suggest that the tool kit is even more applicable in the twenty-first century. Its methods and insights remain powerful. That agency is neither a fiction nor an abstraction; that freedom penetrates far deeper into our being than mere choice; that our nervous system's continuous contact with the material world breathes life into corporeal experience; that such experience is a rich source of qualitative data compatible with the quantitative facts of statistical metrics: these convictions continue to be as pertinent today as they were for the figures in this book.

Our data-driven world is one in which vast numerical spreadsheets impact decision making up and down government, education, and industry. Computer algorithms process Big Data with ever greater computational efficiently and predictive accuracy. Yet, it's worth asking what counts as a datum—particularly in the neurosciences. In 2013, the European Union launched the Human Brain Project, which enlisted the support of a multinational High Performance Analytics and Computing Platform.[5] Researchers contribute medical, cognitive, and statistical information with the goal of generating a "neuroinformatic" atlas of the brain. The output may very well fulfill the dreams originally conjured in the nineteenth century, when quantitative models emerged in mental chronometry and maps of the cerebrum were first drawn up. At that moment and again now, ideas of spirit offer a timely intervention: that our understanding of what the brain does depends as much on what we take selfhood to be. And although any part of our selves can be measured, data are not confined to quantities. Thick data complement thin data. From this vantage we can see that the growth of our knowledge turns not just on the number of data points collected or even on their automation, analysis, or predictive power, but rather on the kinds of human activities that we choose to include in our models. If we could put a spiritualist tool kit forged over the long nineteenth century to use in the twenty-first, it would dilate our data to include stories, contexts, habits, and memories—in short, the spectrum of humans' meaning-making activities.

Yet, our era is not theirs. Over a century has passed since the figures in this book made spirit matter in modern France. What became of their ideas? In the rest of this epilogue, I'd like to trace the ramification of French spiritualism in the history of science and philosophy.

Initially, World War I did not bode well for the spiritualist movement. When hostilities erupted, nationalist fervor engulfed the country's philosophical elite.[6] Soon after Germany and France declared war in August of 1914, Boutroux led the chorus of bellicosity. "Nations in general, and Latin nations in particular, place the essence of civilization in the moral and human life, in the softening and humanizing of customs"; by contrast, he wrote, "according to German thought, softness and kindness are only weakness and powerlessness. Strength alone is strong; and the strength par excellence is science, which, providing us with the strengths of nature, infinitely multiplies our strength."[7] Although he had studied under Germans' tutelage, Boutroux turned his back yet again on the nation that he once fled when the Franco-Prussian war broke out. Bergson also marshaled philosophy in the service of *la patrie*. In 1915, he lambasted Germans for embracing a pernicious *Kultur* premised on "bending science to the satisfaction of material needs."[8] Absorbed as these thinkers were in the war effort, once productive debates between Kantian and spiritualist thought gave way to spiteful antagonism. After the Paris Peace Conference of 1919, when Germany submitted to the Allies' terms, the demobilization of the French troops (and the demobilization of French culture) became all consuming.

As a consequence, the subsequent generation saw ideas of spirit as retrograde. The French Communist Party formed in 1920. Among its membership, the sociologist Georges Friedmann (1902–77) ridiculed spiritualists' idea of the self, which he saw as an "intuition totally abstracted from the world where they lived, taking the real as a breeze of dust."[9] The Hungarian French Georges Politzer (1903–42) decried the quietist bent of those thinkers' preoccupation with psychology, which sought "to comprehend not the individual, but the self [*moi*] in general"; he singled out Bergson for having "always been a zealous ally of the state and of the class for which it is the instrument. Mr. Bergson was openly for the war and in fact against the Russian revolution."[10] Paul Nizan (1905–40) criticized the *guard dogs* of French academia. In his eyes, ideas of spirit had erected a philosophical bulwark against "realist" political engagement.[11] For Léon Blum (1872–1950), the country's socialist prime minister in the decade before the Second World War, spiritualist thought amounted to "anarchic mysticism."[12] A young Michel Foucault (1926–84) dedicated an early essay to French philosophers' century-long preoccupation with psychological theories.[13] Their enormity was not lost on Louis Althusser (1918–90). "From Maine de Biran to Bergson, we can compile, to our dismay, a long list of names: Victor Cousin, Ravaisson, Boutroux, Lachelier, and all their epigones."[14] Rather than intervene in politics, Althusser argued, "It is this tradition that takes up . . . the *religious spiritualism* preserved for us by the Church, its theologians and its ideologues."[15] Ideas of spirit thus came to be seen for some as the bête noire of left politics and for many as an antiscientific dogmatism.

The history of science testifies otherwise. Ideas of spirit had an immediate impact after World War I. Famously, the Collège de France hosted a public debate in 1922 at which Bergson and Alfred Einstein sparred over quantum physics. Intended to build peace between France and Germany via intellectual engagement, the event staged competing epistemologies of temporality: Einstein's mathematical approach clashed with Bergson's psychological account of duration.[16] The public debate kicked off Bergson's productive tenure as the head of the International Committee on Intellectual Cooperation, a subcommittee of the League of Nations that prefigured UNESCO. It was designed to promote an international spirit of collaboration that would stave off war via philosophical and literary dialogue.

More broadly, spiritualist ideas percolated through medical models of selfhood. Psychiatrists brought the strata of inner experience to bear on the classification and treatment of mental illnesses.[17] In *The Troubled Consciousness* (*La conscience morbide*) (1914), Charles Blondel (1876–1939) suggested that the healthy psyche is socialized. It adheres to impersonal, collective time. By contrast, abnormal states "go lost in cenesthopathy, in the mute flux of experience," what he called "pure psychology."[18] Blondel's claim hewed to the spiritualist schema of inner duration and external spatiality. Similarly, Georges Dwelshauvers (1866–1937) contributed to French psychiatry by drawing on the duality of selfhood.[19] It was also influential for the psychiatrist Eugène Minkowski (1885–1972), who elaborated on the temporal bond between subjective experience and social existence.[20] Pathology results from the loss of vital contact with reality. As he saw it, the psychiatrist's task is to diagnose and adjust, by means of intuition and sympathy, patients' personal duration with the world's uniform temporality. Minkowski further contributed etiologies for mental illnesses. Manic depression involves discrete phases of depression and excitation, after which the personality returns to normal; schizophrenia debilitates one's entire temporal sensibility, resulting in the progressive deterioration of emotional capacities. The duality of time and space was also instrumental for Henri Delacroix (1873–1937), who studied the role of language in child development. He showed how the invention of tools—linguistic signs chief among them—was critical for children's minds.[21] These dynamic views of subconscious life were part of a French psychiatric program, which developed independently of Freudian psychoanalysis.[22]

To that end, the work of Pierre Janet (1859–1907) cannot be forgotten. His approach to psychiatry hinged on the principles of dynamism and synthesis. Dynamism described the corporeal activity underlying conscious states. He inferred that maladies such as phobia, hysteria, anxiety, and neurosis should be treated at the level of their motor manifestations.[23] Among the methods that Janet used to draw out patients' subconscious drives were hypnotic suggestion, automatic writing, and the talking cure. Therapy would achieve the goal of synthesizing abnormality into a unified sense of self. He did not only draw on the work of his uncle, the philosopher Paul Janet.[24] Equally influential was the spiritualist premise that psychological activity involves the exertion of embodied force. Janet showed that harmonizing this force with the tensions of mental energy restores the smooth dynamism of the normal psyche.[25]

In biology, Bergson's work has proved impressively durable. Although he and his contemporaries did not live to see significant breakthroughs of the twentieth century—notably, the double-helix structure of DNA—biologists have corroborated central concepts in *Creative Evolution*. Homeotic genes have been shown to facilitate convergent evolution with regard to anatomical structures' development across different species. One such "master control gene" is the *PAX6* gene. It regulates genetic expressions integral to the formation of the eyes (a key case study for Bergson, as I explored in chapter 5) during embryonic development. Genetic coordination begins with light-sensing cells in an enclosed liquid structure and develops into the iris, lens, and retina. Since the completion of the Human Genome Project in 2003, *PAX6* genes have been identified in every animal whose genome has been sequenced. The protein is believed to have existed five hundred million years ago; it predates the divergence of numerous genetic lineages—from shellfish to squid and insects. In a decisive experiment, a mouse *PAX6* was implanted in fruit flies (*Drosophila*). The eye that formed was not a mouse eye, but a fly eye.[26] The process at work was RNA splicing, which controls the expression of the gene sequence found in DNA and generates multiple instructions from the same proteins. It's remarkable how the process had been anticipated by Bergson's insight: "Life does not proceed by the association and addition of elements, but by dissociation and division."[27] Certain components of the RNA (introns) are jettisoned while other components (exons) join together. Consequently, RNA made from the same *PAX6* generates diverse varieties of eyes across species with no common ancestry.

In the neurosciences, spiritualist ideas have proved remarkably formidable. By the 1950s, it became widely accepted that neurotransmitters—and not only electric impulses—are responsible for synaptic communication between neurons. Even though thinkers in this book did not witness the biochemical revolution, subsequent developments in our knowledge of the brain have validated key ideas of theirs. Mass action explains how millions of neurons swarm together in excess of synaptic communications between individual neurons.[28] The critique of associationism has come to prevail among researchers who treat neurons less as the building blocks of cerebral matter and more as moments in a collective and chaotic neural constellation. Additionally, the discovery of neuronal plasticity in the late twentieth century showed how neuronal networks shape conscious experience as much as they are reciprocally shaped by it. Initially, neurons were found to grow in adult birds (namely, canaries and zebra finches) whose songs differed with the seasons owing to shifting control nuclei in their brains.[29] Adult neurogenesis soon became evident in humans as well. Research on cancer patients who received Bromodeoxyuridine (BrdU), a dye used to detect proliferating cells, found that it also marked newborn neurons.[30] The finding overthrew the long-standing belief that neurons could die but not regenerate. Radical reinvention in neural networks is now characteristic of the brain.

These advancements led the French neuroscientist Jean-Pierre Changeux, to remark, "The brain needs to be seen . . . as an open, motivated, and self-organizing system continually engaged in the exploration of its environment—a quality that recalls Bergson's theme in *La Pensée et le mouvant*."[31] Has our understanding of cerebral matter finally become dynamic, heterogeneous, creative—in a nutshell, spiritual? One thing is clear. As much as the contemporary neurosciences have outstripped the knowledge available a century ago, many researchers echo the reductive claims of that period. Consider Joseph LeDoux's evocative words "the essence of who we each are depends on our brains."[32] The spiritualist rejoinder remains apropos as ever.

Beyond the sciences, spirit's influence has been undeniable in the history of philosophy. Initially, the materialist moment in spiritualism predisposed France to the rise of phenomenology thanks to a method predicated on the precise description of the givens of immediate experience and on a holistic metaphysics. Phenomenology was popularized in France following the 1919 arrival of Alexandre Koyré (1892–1964), who introduced Edmund Husserl's (1859–1938) work. Its guiding dictum that philosophers return "back to the things themselves" had a wide impact.[33] In 1928, Max Scheler's (1874–1928) *The Nature of Sympathy* (*Wesen und Formen der Sympathie*) was the first in a line of German books to be translated into French. Emmanuel Levinas (1906–95) followed with the publication of Husserl's lectures.[34] If any single figure thoroughly employed the phenomenological method in tandem with the spiritualist impulse to draw on the physiological and psychological literature of embodiment, it was certainly Maurice Merleau-Ponty (1908–61). He made his debt to the tradition clear in a series of lectures, "The Union of the Soul and the Body in Malebranche, Biran, and Bergson," presented at the École normale supérieure in 1947–48.[35]

In fact, French audiences became familiar with phenomenology in the guise of spirit. Gabriel Marcel (1883–1973) was a Catholic whose brand of spiritualism took the indubitable fact of embodiment as the ground of thought. In his *Metaphysical Journal* (*Journal métaphysique*) (1927) and again in *Being and Having* (*Être et avoir*) (1935), Marcel articulated a phenomenological distinction between two modes in which the body presents itself: as the enfleshed person that I *am* and as the physical capacities that I *have*. Neither description is exhaustive. But they are both consequential. In his plays, Marcel dramatized how one's body can be possessed qua object with the same ease that others can be instrumentalized as mere bodies. In the face of a "broken world," a renewed attunement to our corporeality could make us better respond to the fragile, sensing, and creative mysteries of other finite creatures.[36] Marcel considered the body to be the medium of spirit, an inescapable (and metaphysically pregnant) fact of existence—what Jean Wahl (1888–1974) characterized as part of philosophical orientation toward the *concrete*.[37]

Christians also put ideas of spirit into action across the personalist movement. Jacques Maritain (1882–1973), Emmanuel Mounier (1905–50), and Jean Guitton (1901–99) affirmed the sanctity of the person as a moral ideal, philosophical foundation, and political crusade. The spiritualist view of the self served as their groundwork. For Mounier, personal interiority fueled a relationship with God and the possibility of self-realization. His journal, *Esprit* (founded in 1932), rallied a nonconformist political current, which opposed liberalism and communism for their shared materialism. Both reduced the self to an atomic individual squelched by either market or collectivist forces.[38] For the conservative Maritain, the success of a robust community relied on respect for the person's dignity—a value that he borrowed from Thomism and his early interest in Bergson. The philosopher lent his voice to the personalist cause, claiming, "Each person is protected by an immaterial wall that we call rights, and luckily, we are accustomed today to consider the rights of the person as inviolable."[39] Bergson's final book, *The Two Sources of Morality and Religion* (*Les deux sources de la morale et de la religion*) (1932), was a part of the religious revival in spirit. Following the Second World War, personalist notions proved influential in the reorganization of Europe. At the fledgling United Nations, Maritain's Catholic vision of natural rights steered the Universal Declaration of Human Rights, whose opening line recognized "the inherent dignity" of "all members of the human family."[40]

The "philosophy of spirit," a movement spurred by Louis Lavelle (1883–1951) and René Le Senne (1882–1954), advanced the quest for authenticity amid the alienation of industrial society.[41] For Le Senne, spiritual life involves a continuous dialectic between suffering and overcoming. Suffering prompts one to overcome the finite conditions of existence and to seek the absolute. Much like spirit in the nineteenth century, Le Senne stressed that the will anchored the self. Through volition, one embodies the values of truth, the good, beauty, and love.[42] In a similar key, Lavelle took epistemology to open a pathway to transcendence. In his opus *Dialectic of the Sensible World* (*La dialectique du monde sensible*) (1921) and in *Consciousness of Self* (*La conscience de soi*) (1933), Lavelle took the perception of matter to constitute the data of experience; it indicates what is essential for philosophy: the relation between the finite mind and the infinite realm of spirit. He maintained "that matter is nothing without the spirit that thinks it, that defines the use one must make of it and the signification that one can give to it; after that it is spirit itself which constitutes the true reality, while matter is only the means by which it finds expression."[43] The inherited opposition of spirit and matter remained central. But what counted as material became ossified, a foil against which Lavelle sought to "deepen a certain inner experience without which external experience (that which we have of things) would not be able to hold up."[44] As a result, ideas of spirit took on a new life in the twentieth century, no longer reliant on support from the life and human sciences.

Indeed, the varied itineraries of modern French philosophy illustrate that self-conscious champions of spiritualism were numerous; though their work no longer drew from scientific literature as it had at the turn of the twentieth century. Instead, mystery, being, silence, and ecstasy became guiding motifs among those who espoused fidelity to spirit. Unmoored from lively engagements with neurology and psychology, ideas of spirit came to lead a separate life apart from those of the materialist moment.

That changed in 1966 with Gilles Deleuze's monograph *Bergsonism*, which revitalized the philosopher's standing. According to Deleuze, Bergson's thought was hampered by neither psychologism nor irrationalism. As Deleuze read him, Bergson's intuitive method lent analytic rigor to his statement and creation of problems. Most important among them was the relation between science and metaphysics. When Deleuze returned to write an epilogue to his book two decades after its initial publication, he revisited the central problem of Bergson's work: "Bergson did not merely criticize science as if it went no further than space, the solid, the immobile. Rather, he thought that the Absolute has two 'halves,' to which science and metaphysics correspond. Thought divides into two paths in a single impetus, one toward matter, its bodies and movements, and the other toward spirit, its qualities and changes."[45]

In returning to the duality of spirit matter, Deleuze made Bergson a thinker to be reckoned with—and not relegated to the history section of the lycée philosophy course. Vibrant scholarship has followed around Bergson's corpus. Bergson continues to enjoy recognition as a robust thinker beyond just philosophy—including feminism,[46] aesthetics,[47] racial identification,[48] religion,[49] and human rights.[50] Yet, Bergson's renewed fame has come at a price. Ahistorical accounts are still written about a towering individual thinker plucked from the intellectual and institutional contexts that had sustained his writing. Alarmingly, some scholars even take *l'esprit* to be a dirty word. But wresting Bergson from the spiritualist heritage leaves us with an impoverished understanding of the stakes that lent significance to his work. The problems motivating his philosophy were not exclusively philosophical. They were also scientific. And although Bergson rose to the apex of the French academic establishment, his concerns were not so rarefied. By returning to the literature that he read, the societies in which he participated, and the classrooms in which he taught, my aim has been to show how Bergson contributed to a constellation of thought that instilled a critical sensibility in peers and students alike. Their vision of the self did not have to choose between the humanities and sciences; it was rooted in a synthesis of historical comprehension, scientific precision, and social engagement.

The figures in this book shared a shrewd appreciation for the advancements of the sciences as well as the aporias in the sciences. In their approaches to scientific literature, spiritualist thinkers in nineteenth- and twentieth-century France made a point to show how brain research is not exclusively scientific; a conceptual surplus remains in excess of its findings. If there's a moral to the materialist moment in spirit, it is that the sciences yield interpretive questions that resonate beyond empirical inquiry and supply renewed traction to philosophical, religious, political, and anthropological methods. Our evolving understanding of the brain offers a launching pad from which to imagine new problems—and possibilities—for ourselves.

Acknowledgments

There is a bookstore in Berkeley called Moe's Books. I have visited intermittently since my years as an undergraduate student nearby at the University of California, returning to thumb through pages and to buy whatever my bank account would allow. The origins of *Making Spirit Matter* can be traced to my wanderings about the philosophy section on the third floor—specifically, to the day that I purchased Henri Bergson's *Matter and Memory,* which Daniel Coffeen had assigned in his rhetoric course. No other book has proved so auspicious. My efforts to understand Bergson set in motion a decade-long journey exploring the history of the brain sciences in France. But origins explain only so much. The claims made in the preceding chapters gradually took shape over the course of recurrent visits to Moe's. Armed with the suggestions of teachers, friends, and itinerant interlocutors, I browsed sections beyond philosophy. New worlds opened themselves; my assumptions crumbled apart. After all, recommending a book is no trivial matter. I owe a debt of gratitude to the following people for having done so.

The history section on the fourth floor captivated me thanks to my doctoral advisers at Johns Hopkins University's Humanities Center. Above all, Ruth Leys has been an intellectual beacon guiding my scholarship through the problems at stake in the human sciences. Her attunement to argument, contradiction, and disagreement offered a model of scholarly rigor. Gabrielle Spiegel opened my eyes to methodological debates in historiography. She has been an enthusiastic supporter well beyond my time in Baltimore. Paola Marrati has been a companion on both sides of the Atlantic. I am fortunate to have read Bergson's oeuvre with someone of such uncompromising philosophical precision. Warren Breckman welcomed me to the University of Pennsylvania. At the center of his modern European intellectual history seminar was a robust eclecticism that brought disparate strands of thought into dialogue. As a result, I would return to Moe's Books during the holidays with an ever-weightier reading list.

That list expanded thanks to others in Baltimore. Hent de Vries was a masterful teacher of the history of philosophy, assiduously demonstrating metaphysics' intractability. Tarek Dika reminded me never to claim victory too easily. Our arguments are only as convincing as the opponents we target. Numerous teachers and friends were sources of inspiration: Jane Bennett, William Connolly, Veena Das, Loumia Ferhat, Jonathon Hricko, Nicole Jerr, Jessica Lamont, Leonardo Lisi, Omid Mehrgan, Yitzhak Melamed, Anne Moss, Nils Schott, Todd Shepard, Martin Shuster, and Jennifer Watson. I would be remiss not to mention Michael Fried, whose insistence that I read Maine de Biran and Félix Ravaisson was not inconsequential.

As it turned out, Moe's carried neither. The French section was limited. So, I went to Paris.

There, Frédéric Worms hospitably received me into a community of Bergsoniennes while I carried out studies at the École normale supérieure. Elie During opened my eyes to the institutions and idiosyncrasies of French society. Giuseppe Bianco and Jacob Levi showed me how to enjoy them. My archival research benefited from numerous librarians in France, curious as they were that an American would be so interested in *désuet* thinkers relegated to the waste bins of the past. These people painstakingly brushed the dust off old spiritualists' papers: Françoise Dauphragne of the École normale supérieure; Mireille Pastoureau, conservateur général of the Bibliothèque de l'Institut de France; Serge Sollogoub of the Institut Catholique de Paris; Karine Faren of the Université Paul Valéry—Montpellier III. Particularly charitable was Jean Leclercq of the Université Catholique de Louvain, who opened up Maurice Blondel's personal library to me. Thanks to their assistance, I unearthed a community of dead thinkers who—although distant—became a familiar part of my life.

Back at Moe's Books, the anthropology and science and technology studies sections came to preoccupy me. Suggestions poured in from my colleagues at Wesleyan University, where I wrote the bulk of the preceding chapters. I am immensely thankful to Ethan Kleinberg for having invited me to Middletown to explore materialism with a cadre of fellows in the Center for the Humanities. Appreciation is also extended to my colleagues Hassan Almohammed, Lori Gruen, Steven Horst, Cecilia Miller, Jill Morawski, Victoria Pitts-Taylor, Michael Roth, Joseph Rouse, Jeanette Samyn, Matthew Specter, and Kari Weil. I wonder how this book might have ended up without their insights? Likely in a far worse state.

The same can be said for the incisive comments of those who read drafts, including Michael Behrent, Isabel Gabel, Stefanos Geroulanos, and Sam

Moyn. Jennifer Ratner-Rosenhagen entertained the myriad titles (and wisecracks) that I sent her way. Musings with Francesca Bordogna, Dani Holtz, Emily Marker, and Christopher Nichols lent texture to the preceding pages' prose. Steven Vincent has not just been a kind friend; his vigorous defense of French neo-Kantianism has served as a reminder that intellectual history is a field of reasoned criticism.

The spiritual labor of writing is impossible without the material support of institutions maintained by diligent human beings. For that I owe much to the administrative coordinators Marva Philip and Erinn Savage. Their tireless (and often thankless) work at Johns Hopkins University and Wesleyan University, respectively, facilitated unruffled environments in which creativity thrives. Moreover, this book would not have been written without the financial support provided by the Andrew W. Mellon Foundation, the National Science Foundation, and the Richard A. Macksey Fund. In addition, an earlier version of chapter 3 was published as "Confronting the Brain in the Classroom: Lycée Policy and Pedagogy in France, 1874–1902," *History of the Human Sciences* 28, no. 1 (2015): 3–24.

My editors at the University of Chicago Press were crucial. From the beginning, Douglas Mitchell gushed with excitement about my endeavor to bridge the two cultures of the humanities and sciences. The memory of his joyous enthusiasm beats to the vibrant rhythm of a drum solo. Priya Nelson contributed her keen sense of narrative. Dylan Montanari offered skillful support along the way.

Creative work is a continuous labor that begins over and over again. Thanks to exchanges, arguments, idle banter, and pontification with close friends, I began this book over and over again—reworking the arguments and kneading the sentences. These thoughtful people include Shikha Bhattacharjee, Scott Browne, Tony Chiarito, Petey Gil-Montllor, Bryce Goodman, Hirsh Jain, Matt Kelly, Will Leiter, Adwait Parker, Ryan Patterson, and Dan Sheehan. Claudia Biçen perceived well before I did the deep connections between science and spirit. My parents, Carolyn and Larry Craig McGrath, inculcated a work ethic that fueled this project. Rachel Bobrick gave me the energy to bring it to completion. And my sister, April, imparted a sense of humor, which (I hope) occasionally shines through.

My most recent trip to Moe's Books was spent dwelling in the religion section. Fittingly, I returned to the opposite side of the same floor that housed philosophy. I picked up two books, which I hoped would unravel the Jewish and Christian facets of Bergson's thought. As I flipped through the pages, memories flashed of those who instilled a sensibility for writing and reading. At UC Berkeley, Nancy Weston showed me how there is still more to be thought than we are accustomed to supposing. Pheng

Cheah and I read Nietzsche closely together. Hubert Dreyfus taught me how to cope. My memories plunged deeper; the planes of consciousness expanded. I recalled Eileen Parker and Louise McFadden, who had nurtured an unrelenting curiosity during high school. I bought the books, put them in my bag, walked down Telegraph Avenue, and reflected on how it was that the same bookstore revealed shifting shades of significance. Of course, Moe's Books didn't change. I did: the self wandering the shelves.

Archives Consulted

Archives nationales, Paris

Fonds École normale supérieure
Ministère de l'instruction publique

Bibliothèque de l'Institut de France, Paris

Cahiers d'étudiant et papiers divers d'Émile Boutroux
Fonds Louis Hautecoeur
Notules et fragments de la correspondance de Raymond Thamin
Opuscules et correspondance de Jules Lachelier
Papiers de Charles Lévêque

Bibliothèque littéraire Jacques Doucet, Paris

Fonds Henri Bergson

Bibliothèque nationale de France, Paris

Fonds Henri Lachelier
Fonds Marcel Proust
Fonds Théodule Ribot
Papiers d'Alain

Bibliothèque Victor Cousin, la Sorbonne, Paris

Papiers Victor Egger
Papiers Henri Bergson

École normale supérieure, Paris

Manuscrits Émile Boutroux

Institut Catholique de Paris

Fonds Édouard Le Roy

Université Catholique de Louvain, Louvain-La-Neuve, Belgium

Archives de Maurice Blondel

Université Paul Valéry—Montpellier III

Fonds Charles Renouvier

Notes

Introduction

1. See Harry W. Paul, *From Knowledge to Power: The Rise of the Science Empire in France, 1860–1939* (New York: Cambridge University Press, 2003).

2. "Spiritualism" is distinct from "spiritism," understood as contact with another spectral world. Spiritism drew wide interest in France during the late nineteenth century, but it is not the subject of this book. See M. Brady Bower, *Unruly Spirits: The Science of Psychic Phenomena in Modern France* (Champaign: University of Illinois Press, 2010); John Warne Monroe, *Laboratories of Faith: Mesmerism, Spiritism, and Occultism in Modern France* (Ithaca, NY: Cornell University Press, 2007).

3. Jean Louis Fabiani, *Qu'est-ce qu'un philosophe français? La vie sociale des concepts (1880–1980)* (Paris: Éditions de l'École des hautes études en sciences sociales, 2010), 153. Translations from the original French are my own unless otherwise noted.

4. Louis Racine, *Religion, a Poem* (London: W. Strahan, 1754), 27. "Je pense. La pensée, éclatante lumière, / Ne peut sortir du sein de l'épaisse matière. / J'entrevois ma grandeur. Ce corps lourd et grossier / N'est donc pas tout mon bien, n'est pas moi tout entier."

5. Bergson, "Psychophysical Parallelism and Positive Metaphysics," in *Continental Philosophy of Science*, trans. Jean Gayon, ed. Gary Gutting (Oxford: Blackwell, 2005), 63–64; originally published as "Séance du 2 mai 1901: Le Parallélisme psychophysique et la métaphysique positive," *Bulletin de la Société française de philosophie* 1 (1901): 33–71.

6. Ibid., 63.

7. See Gerald Izenberg, *Identity: The Necessity of a Modern Idea* (Philadelphia: University of Pennsylvania Press, 2016); Charly Coleman, *The Virtues of Abandon: An Anti-individualist History of the French Enlightenment* (Stanford, CA: Stanford University Press, 2014); Raymond Martin and John Barresi, *The Rise and Fall of Soul and Self* (New York: Columbia University Press, 2006); Jerrold Seigel, *The Idea of the Self: Thought and Experience in Western Europe since the Seventeenth Century* (New York: Cambridge University Press, 2005); Jan Goldstein, *The Post-Revolutionary Self: Politics and Psyche in France, 1750–1850* (Cambridge, MA: Harvard University Press, 2005); Dror Wahrman, *The Making of the Modern Self: Identity and Culture in Eighteenth-Century England* (New Haven, CT: Yale University Press, 2004); Nikolas Rose, *Inventing Our Selves: Psychology, Power, and Personhood* (Cambridge: Cambridge University Press, 1998).

8. Interiority emerged as a modern feature of the self in contrast to the medieval

regime of selfhood, for which external categories such as dress, rank, or guild were definitive of the person. See Roy Porter, ed., *Rewriting the Self: Histories from the Middle Ages to the Present* (New York: Routledge, 1997). Marcel Mauss traced this deepening conceptual trajectory beginning with the self as a superficial designation and culminating in the private domain of inner personhood in "A Category of the Human Mind: The Notion of Person, the Notion of 'Self'" (1938) in *Sociology and Psychology: Essays*, ed. Ben Brewster (London: Routledge and Kegan Paul, 1979), 61. Similarly, Michel Foucault presents interiority as a relatively recent invention of modernity in *The Order of Things: An Archaeology of the Human Sciences*, 1966 (New York: Vintage, 1994).

9. Charles Taylor, *Sources of the Self* (Cambridge, MA: Harvard University Press, 1989), 185–93.

10. Blaise Pascal, *The Thoughts, Letters, and Opuscules of Blaise Pascal*, trans. O. W. Wight (New York: Hurd and Houghton, 1869), 26.

11. Jean-Jacques Rousseau, *A Discourse on the Origins of Inequality*, 1755, trans. Maurice Cranston (New York: Penguin Books, 1984), 88.

12. Claude-Adrien Helvétius, *De l'esprit* (Paris: Durand, 1758), 7.

13. Pierre Jean-George Cabanis, *On the Relations between the Physical and Moral Aspects of Man*, vol. 1, trans. Margaret Duggan Saidi (Baltimore: Johns Hopkins University Press, 1981), 33; originally published as *Rapport du physique et du moral de l'homme*, vol. 1 (Paris: Crapelet, 1805).

14. Much of the literature addressing French spiritualism has organized the tradition into a filiation spanning the nineteenth century, from Maine de Biran to Henri Bergson. One of my arguments is that this trajectory was by no means linear. Biran enjoyed a belated reception in the late nineteenth century as the ex post facto origin of the materialist moment in spiritualism. See Andrea Bellantone, "Coup d'œil sur Bergson et le spiritualisme français," in *Actualité d'Henri Bergson*, ed. Tamás Ullmann and Jean-Louis Vieillard-Baron (Paris: Archives Karéline, 2012), 85–100; F. C. T. Moore, *The Psychology of Maine de Biran* (London: Cambridge University Press, 1966); George Boas, "Bergson and His Predecessors," *Journal of the History of Ideas* 20, no. 4 (1959): 503–14; Ben-Ami Scharfstein, *Roots of Bergson's Philosophy* (New York: Columbia University Press, 1943); Susan Stebbins, *Pragmatism and French Voluntarism, with Especial Reference to the Notion of Truth in the Development of French Philosophy from Maine de Biran to Bergson* (London: Cambridge University Press, 1914). Gabriel Madinier offers the most comprehensive account of nineteenth-century spiritualism in *Conscience et mouvement, étude sur la philosophie française de Condillac à Bergson* (Paris: Presses universitaires de France, 1938).

15. See Joan W. Scott, "Against Eclecticism," *Differences: A Journal of Feminist Cultural Studies* 16, no. 3 (2005): 114–37; Donald Kelley, *The Descent of Ideas: The History of Intellectual History* (Burlington, VT: Ashgate, 2002), 9; Kelley, "Eclecticism and the History of Ideas," *Journal of the History of Ideas* 62, no. 4 (2001): 577–59; John L. Brooks III, *The Eclectic Legacy: Academic Philosophy and the Human Sciences in Nineteenth-Century France* (Newark: University of Delaware Press, 1998).

16. Goldstein, *Post-Revolutionary Self*, 6.

17. Goldstein argues that Cousin's influence endured in French philosophy well into the twentieth century in "Neutralizing Freud: The Lycée Philosophy Class and the Problem of the Reception of Psychoanalysis in France," *Critical Inquiry* 40, no. 1 (2013): 40–82.

18. Étienne Vacherot, *Le nouveau spiritualisme* (Paris: Hachette, 1884), iv.

19. Ibid., 255.

20. Jules Lachelier, "Cours de Psychologie par M. Lachelier, 1866–67." MS 4118.2. Cahiers d'étudiant et papiers divers d'Émile Boutroux, Bibliothèque de l'Institut de France, Paris.

21. Jules Lachelier, letter to Émile Boutroux, September 13, 1870, MS 4687, Opuscules et correspondance de Jules Lachelier, Bibliothèque de l'Institut de France, Paris.

22. Jules Lachelier, letter to Émile Boutroux, April 2, 1871, ibid.

23. Eric Hobsbawm had suggested, "the era from 1870 to 1914 was above all, in most European countries, the age of school." *The Age of Empire, 1875–1914* (New York: Pantheon Books, 1987), 150.

24. George Weisz illustrates the failures of universities to achieve Republicans' dream of social integration in *The Emergence of Modern Universities in France, 1863–1914* (Princeton, NJ: Princeton University Press, 1983).

25. Émile Boutroux, *The Contingency of the Laws of Nature*, trans. Fred Rothwell (Chicago: Open Court, 1911), 146; originally published as *De la contingence des lois de la nature* (Paris: Germer Baillière, 1874).

26. Jules Lachelier, *Du fondement de l'induction* (Paris: Librairie philosophique de Ladrange, 1871), 102.

27. Gustave Belot, "Un nouveau spiritualisme," *Revue philosophique* 44 (1897): 182–99.

28. Paul Janet, "Une nouvelle phase de la philosophie spiritualiste," *Revue des deux mondes* 108 (1873): 370.

29. Vacherot, *Le nouveau spiritualisme*, i.

30. B. Jacob, "La philosophie d'hier et celle d'aujourd'hui," *Revue de métaphysique et de morale* 6, no. 2 (1898): 177.

31. Henri Bergson, *Creative Evolution*, trans. Arthur Mitchell (New York: Henry Holt, 1911), 268; originally published as *L'évolution créatrice* (Paris: Félix Alcan, 1907).

32. Caterina Zanfi, *Bergson et la philosophie allemande: 1907–1932* (Paris: Armand Colin, 2013); Larry S. McGrath, "Bergson Comes to America," *Journal of the History of Ideas* 74, no. 4 (2013): 599–620; Tom Quirk, *Bergson and American Culture* (Chapel Hill: University of North Carolina Press, 1990); Paul Douglass, *Bergson, Eliot, and American Literature* (Lexington: University of Kentucky Press, 1986); Anthony E. Pilkington, *Bergson and His Influence: A Reassessment* (Cambridge: Cambridge University Press, 1976).

33. Arnaud François and Camille Riquier, eds., *Annales bergsoniennes, VI: Bergson, le Japon, la catastrophe* (Paris: Presses universitaires de France, 2013); Souleymane Bachir Diagne, *Bergson postcolonial: L'élan vital dans la pensée de Léopold Sédar Senghor et de Mohamed Iqbal* (Paris: CNRS, 2011); *African Art as Philosophy: Senghor, Bergson and the Idea of Negritude*, trans. Chike Jeffers (Chicago: University of Chicago Press, 2011); Hilary L. Fink, *Bergson and Russian Modernism, 1900–1930* (Evanston, IL: Northwestern University Press, 1999); Enrique Dussel, "Philosophy in Latin America in the Twentieth Century: Problems and Currents," in Eduardo Mendieta, ed., *Latin American Philosophy: Currents, Issues, Debates* (Bloomington: Indiana University Press, 2003), 11–53; Alain Guy, "Le bergsonisme et Amérique latine," *Caravelle* 1, no. 1 (1963): 121–39.

34. Dominique Janicaud, *Ravaisson et la métaphysique: Une généalogie du spiritualisme français* (Paris: Vrin, 1998), 2. Jean-Michel Salanskis similarly organizes the

spiritualist tradition in *Le concret et l'idéal: Levinas vivant III* (Paris: Klincksieck, 2015). Moreover, Gary Gutting recognizes that spiritualists in the late nineteenth century "saw their reflections as grounded in an accurate understanding and appreciation of scientific results." *French Philosophy in the Twentieth Century* (Cambridge: Cambridge University Press, 2001), 8.

35. Carl Vogt, *Physiologische briefe für gebildete aller stände*, 2nd ed. (Geißen: Richer, 1853), 322.

36. Karl Marx, *The Eighteenth Brumaire of Louis Bonaparte*, 1852, trans. Saul K. Padover, https://www.marxists.org/archive/marx/works/1852/18th-brumaire/index.htm (accessed May 5, 2019).

37. James Kloppenberg traces a strand of social thought from the same period that developed out of a *via media* between metaphysics and science. *Uncertain Victory: Social Democracy and Progressivism in European and American Thought, 1870–1920* (New York: Oxford University Press, 1988).

38. Among the proponents of spiritualist materialism, we find the progenitors of new materialisms, which claim to unlock emancipatory theories of affect, vitality, and embodiment. See Catherine Keller and Mary-Jane Rubenstein, eds., *Entangled Worlds: Religion, Science, and New Materialisms* (New York: Fordham University Press, 2017); Victoria Pitts-Taylor, ed. *Mattering: Feminism, Science, and Materialism* (New York: New York University Press, 2016); Brian Massumi, *The Politics of Affect* (New York: Polity, 2015); William E. Connolly, *A World of Becoming* (Durham, NC: Duke University Press, 2011); Jane Bennett, *Vibrant Matter: A Political Ecology of Things* (Durham, NC: Duke University Press, 2010); Diane Coole and Samantha Frost, eds., *New Materialisms: Ontology, Agency, and Politics* (Durham, NC: Duke University Press, 2010); Elizabeth Wilson, *Psychosomatic: Feminism and the Neurological Body* (Durham, NC: Duke University Press, 2004). Notwithstanding their political claims, these and other proponents of new materialisms would stand to benefit from revisiting spiritualist materialism, which countenanced matter's productivity, yet eschewed its autonomy from human thought and action. The thinkers in this intellectual formation took the generative powers of matter also to involve its conceptual features.

39. Eugen Weber characteristically writes, "The advocates of reason and science thought it possible to explain and rule the world. . . . But a reaction soon developed, one that stressed irrational factors such as the unknown, the mysterious, and the wonderful." *France, Fin de Siècle* (Cambridge, MA: Belknap Press of Harvard University Press, 1986), 142. Weber's narrative hewed to Carl Schorske's in *Fin-de-Siècle Vienna: Politics and Culture* (New York: Vintage, 1981).

40. Hughes uses the term "positivism," as he avers, "to characterize the whole tendency to discuss human behavior in terms of analogies drawn from natural science." H. Stuart Hughes, *Consciousness and Society: The Reorientation of European Social Thought, 1890–1930* (New York: Vintage, 1958), 37. Yet, as W. M. Simon argues, positivism, in the strict sense of Comte's "conception of the world and of man," hardly found academic adherents by the 1890s—the generation that, according to Hughes, led the revolt. W. M. Simon, *European Positivism in the Nineteenth Century* (Ithaca, NY: Cornell University Press, 1963).

41. R. C. Gronin depicts the spiritualist period as a religious revival in *The Bergsonian Controversy in France, 1900–1914* (Calgary: University of Calgary Press, 1988), 1–20.

42. J. W. Burrow, *The Crisis of Reason: European Thought, 1848–1914* (New Haven, CT: Yale University Press, 2000), 56–67.

43. James A Winders, *European Culture since 1848: From Modern to Postmodern and Beyond* (New York: Palgrave, 2001), 94.

44. Thomas Laqueur, "Why the Margins Matter: Occultism and the Making of Modernity," *Modern Intellectual History* 3, no. 1 (2006): 112.

45. Félix Ravaisson, *Rapport sur la philosophie en France au XIXe siècle: Recueil de "Rapports" sur les progrès des lettres et des sciences en France* (Paris: Hachette, 1867), 32.

46. Bergson, "Psychophysical Parallelism and Positive Metaphysics," 68.

47. Édouard Le Roy, "Un positivisme nouveau," *Revue de métaphysique et de morale* 9, no. 2 (1901): 139.

48. Frédéric Paulhan, *La physiologie de l'esprit* (Paris: Félix Alcan, 1880), 6.

49. Today, the autonomic nervous system is divided into sympathetic functions, responsible for muscular activation, and parasympathetic functions, responsible for the conservation of bodily resources.

50. Ruth Leys, *From Sympathy to Reflex: Marshall Hall and His Opponents* (New York: Garland, 1990), 18.

51. William Benjamin Carpenter, *Principles of Human Physiology* (London: J and A Churchill, 1846), 515.

52. Volition (*volonté*) is an effortful corporeal activity, which is distinct from the disembodied decision involved in *free will* (in French, *libre arbitre*). The latter concept developed from Saint Augustine, who had conceived *liberum arbitrium*, or "free decision" as the highest form of freedom. See Augustine, *On Free Choice of the Will*, trans. Thomas Williams (Indianapolis: Hackett, 1993).

53. My approach builds on that of Frédéric Worms, who identifies the "problem of spirit" as the grounding moment of twentieth-century French philosophy. Worms contends that spirit was not opposed to another physical, material, or natural dimension of reality. It was excavated from *within* the sciences of the time. See *La philosophie en France au XXe siècle* (Paris: Gallimard, 2009), 31–64.

54. The field of nineteenth-century French philosophy has often been framed in a threefold scheme of spiritualism, idealism, and positivism. Dominique Parodi originally employed the frame in *La philosophie contemporaine de la France: Essai de classification des doctrines* (Paris: Félix Alcan, 1919); a similar threefold distinction is also adopted by Jean Guitton, who charts spiritualism, idealism, and rationalism in *Regards sur la pensée française: 1870–1940* (Paris: Beauchesne, 1968). Isaac Benrubi identified a more nuanced cut across the three "great currents" of French philosophy at the turn of the century: (1) an empiricist and scientific positivism associated with Ernst Renan, Émile Littré, and Hippolyte Taine; (2) a critical and epistemological idealism promoted by neo-Kantians, above all Charles Renouvier; and (3) a "metaphysical and spiritualist positivism." *Les courants de la philosophie contemporaine en France* (Paris: Félix Alcan, 1933), 4. Also see Christian Dupont, *Phenomenology in French Philosophy: Early Encounters* (New York: Springer, 2014), 22–37; F. C. T. Moore, "French Spiritualist Philosophy," in *Nineteenth-Century Philosophy: Revolutionary Responses to the Existing Order*, ed. Alan D. Schrift and Daniel Conway (New York: Routledge, 2014), 161–76; Jean Beaufret, *Notes sur la philosophie en France au XIXe siècle: De Biran à Bergson* (Paris: Vrin, 1984); Jean Theau, *La philosophie française dans la première moitié du XXe siècle* (Ottawa: Éditions d'Université d'Ottawa, 1977).

55. Michel Foucault, "Life: Experience and Science," in *Essential Works of Fou-*

cault, 1954–1984, vol. 2, *Aesthetics, Method and Epistemology*, ed. James D. Faubion, trans. Robert Hurley and Paul Rabinow (New York: New Press, 1998), 466; originally published as "La vie: L'expérience et la science," *Revue de métaphysique et de morale* 70 (1985): 4.

56. Knox Peden arranges twentieth-century French philosophy along similar lines in *Spinoza contra Phenomenology* (Palo Alto, CA: Stanford University Press, 2014); Elie During contests Foucault's lineage, preferring to place Bergson at the intersection of the two traditions, in "'A History of Problems': Bergson and the French Epistemological Tradition," *Journal of the British Society for Phenomenology* 35, no. 1 (2004): 4–23.

Chapter One

1. See Christophe Prochasson, *Les années électriques, 1880–1910* (Paris: La Découverte, 1991), 85.

2. "The spirit nourishes within," from Virgil's *Aeneid*, VI, 726. See *The Aeneid*, ed. Vincent J. Cleary (Ann Arbor: University of Michigan Press, 1995), 142.

3. Hippolyte Taine, *Les philosophes classiques du XIXe siècle*, 3rd. ed. (Paris: Hachette, 1868), vi.

4. Ibid., viii.

5. Ravaisson, *Rapport sur la philosophie en France au XIXe siècle*, 258, my emphasis. On the significance of Ravaisson's report, see François Azouvi, *La gloire de Bergson: Essai sur le magistère philosophique* (Paris: Gallimard, 2007), 27; Henri Gouhier, *Bergson et le christ des évangiles* (Paris: Arthème Fayard, 1961), 45–46.

6. Ravaisson, *Rapport sur la philosophie en France au XIXe siècle*, 261.

7. Henri Bergson, "The Life and Work of Ravaisson," 1904, in *The Creative Mind: An Introduction to Metaphysics*, trans. Mabelle L. Andison (Mineola, NY: Dover, 2007), 204; *The Creative Mind* originally published as *La pensée et le mouvant* (Paris: Félix Alcan, 1934).

8. Ibid.

9. Victor Cousin, *Fragments philosophiques* (Paris: Ladrange, 1838), 397.

10. Victor Cousin, "Avant-Propos de l'édition de 1853," *Du vrai, du beau, du bien*, 9th ed. (Paris: Didier, 1862), vii.

11. Kelley, *Descent of Ideas*, 9.

12. Goldstein, *Post-Revolutionary Self*, 183.

13. Ravaisson, *Rapport sur la philosophie en France au XIXe siècle*, 232.

14. Félix Ravaisson's many government reports include: *Rapport adressé à s. exc. le ministre d'état au nom de la commission* (Paris: Typographie E. Panckoucke, 1862); *De l'enseignement du dessin dans les lycées* (Paris: P. Dupont, 1853); *Catalogue général des manuscrits des bibliothèques publiques des départements, publié sous les auspices du ministre de l'instruction publique* (Paris: Imprimerie Royale, 1849); *Rapports au ministre de l'instruction publique sur les bibliothèques des départements de l'ouest* (Paris: Joubert, 1841).

15. Benjamin Aubé, "Compte rendu," *Revue de l'instruction publique*, July 30 1868, 292–94.

16. Victor Duruy, "Séance du 23 mai 1868," in *L'administration de l'instruction publique de 1863 à 1869* (Paris: Jules Delalain, 1869), 627.

17. Ibid., 629.

18. For a decade under the Second Empire, secondary school students took a course in logic instead of philosophy. Doctoral candidates could pursue a mollified degree in rhetoric. See Bruno Poucet, "Comment s'élaborent les contenus de programme en philosophe de 1863 à 1890?," *Spirale—revue de recherches en éducation* 14 (1995): 59–102.

19. Jacques Derrida read in Ravaisson's report a turn away from the primacy of perception that lay at the foundation of Enlightenment epistemology. With Ravaisson, the haptic overtook the optic, what Derrida called a "history of the body." *On Touching—Jean-Luc Nancy*, 2000, trans. Christine Irizarry (Stanford, CA: Stanford University Press, 2005), 137.

20. Jules Lachelier, *Revue de l'instruction publique*, September 10, 1868, 386–87.

21. Jules Lachelier, letter to Félix Ravaisson, August 15, 1868, MS 4119, Cahiers d'étudiant et papiers divers d'Émile Boutroux, Bibliothèque de l'Institut de France, Paris.

22. François Magy, *La raison et l'âme: Principes du spiritualisme* (Paris: A. Durand et Pedone-Lauriel, 1877), 187.

23. Fabien Capeillères argues that Ravaisson, Lachelier, and Boutroux articulated spiritualist positivism while navigating neo-Kantian conceptions of the positive sciences. See Capeillères, "To Reach for Metaphysics: Émile Boutroux's Philosophy of Science," in *Neo-Kantianism in Contemporary Perspective*, ed. Rudolf A. Makkreel and Sebastian Luft (Bloomington: Indiana University Press, 2010), 192–252. These thinkers belonged to a wave of French intellectuals who studied idealist philosophy in Germany. See Michel Espagne, *En deçà du Rhin: L'allemagne des philosophes français au XIXe siècle* (Paris: Éditions du Cerf, 2004).

24. Henri Gouhier, *Les conversions de Maine de Biran* (Paris: J. Vrin, 1948), 226.

25. Maine de Biran, *Maine de Biran, sa vie et ses pensées*, ed. Ernest Naville (Paris: Joel Cherbuliez, 1857), 243–44.

26. See Maurice Merleau-Ponty, *The Incarnate Subject: Malebranche, Biran, and Bergson on the Union of Body and Soul*, 1948, trans. Paul B. Bilman (Amherst, NY: Humanity Books, 2001).

27. Biran originally submitted the treatise in 1799 in response to the topic: "Determine what is the influence of habit on the faculty of thought; or, in other words, show the effects made on our intellectual faculties by the frequent repetition of their very operation." His other works to reach the press during his lifetime were a minor essay on the philosophy of Pierre Laromiguière and another on Leibniz. See Maine de Biran, "Exposition de la doctrine philosophique de Leibniz," in *Biographie Universelle*, vol. 23 (Paris: Michaud, 1819); *Examen des leçons de philosophie de M. Laromiguière* (Paris: Fournier, 1817).

28. Paul Janet, *Les problèmes du XIXe siècle* (Paris: Michel Lévy Frères, 1872), 284.

29. See Martin S. Staum, *Minerva's Message: Stabilizing the French Revolution* (Montréal: McGill-Queen's University Press, 1996), 95–117.

30. Cited in Gouhier, *Les conversions de Maine de Biran*, 136–54.

31. Maine de Biran, *Influence de l'habitude sur la faculté de penser*, ed. Pierre Tisserand (Paris: Presses universitaires de France, 1953), 87. The physiologist Xavier Bichat also formulated a version of the double law of habit: "The feeling is constantly blunted by it, whereas the judgment on the contrary owes to it its perfection." See *Physiological Researches upon Life and Death*, 1800, trans. Tobias Watkins (Philadelphia: Smith and Maxwell, 1809), 24; originally published as *Recherches physiologiques sur la vie et la mort* (Paris: Brosson, 1800).

32. This monistic understanding of active habits hewed to Bichat's claim, "the center of those revolutions of pleasure, of pain, and of indifference is not in the organs which receive or transmit the sensation, but in the mind which perceive it." Bichat, *Physiological Researches*, 37.

33. Maine de Biran, *Influence de l'habitude sur la faculté de penser*, 75.

34. Ibid., 55.

35. Henri Gouhier, "Introduction," in *Maine de Biran, œuvres choisies* (Paris: Aubier-Montaigne, 1942), 31–32.

36. Maine de Biran, *Influence de l'habitude sur la faculté de penser*, 55.

37. Michel Henry argues that Biran introduced the first phenomenology of the body through his idea of the immanence of experience. See Henry, *Philosophie et phénoménologie du corps* (Paris: Presses universitaires de France, 1965).

38. See Elizabeth A. Williams, *The Physical and the Moral: Anthropology, Physiology, and Philosophical Medicine in France, 1750–1850* (New York: Cambridge University Press, 1994), 85.

39. Cabanis, *On the Relations between the Physical and Moral Aspects of Man*, 1:33.

40. Ibid., 94–95.

41. Ibid., 95.

42. Ibid., 97.

43. Biran rejected the crudely materialist implications of Cabanis's physiology, which treated the brain "as a peculiar organ, specially designed for the production [of ideas], just as the stomach is designed to effect digestion, the liver to filter bile, the parotids and the maxillary and sublingual glands to prepare the salivary juices." See ibid., 152–53. This view did influence the first draft of Biran's treatise on habit. But in the published edition, Biran parted from Cabanis's belief that consciousness comprises bodily motions alone. See Pierre Tisserand, "La fécondité des idées philosophiques de Maine de Biran," *Bulletin de la Société française de philosophie* 24, nos. 2–3 (1924): 29.

44. Bichat, *Physiological Researches*, 3.

45. Ibid.

46. Ibid.

47. Maine de Biran, *Influence de l'habitude sur la faculté de penser*, 226.

48. The primary goal of the society was to produce a medical cartography of the region. The members conducted health surveys and documented diseases, agriculture, potable water sources, and livestock. Biran saw to it that mayors and pastors of the predominantly Protestant region were advised about the benefits of vaccination, a notable endeavor in an age prior to the discovery of bacteria. See Paul Marx, "Maine de Biran (1766–1824), fondateur de la société médicale de Bergerac," *Histoire des sciences médicales* 32, no. 4 (1998): 285–388; Pierre Lemay, *Maine de Biran et la société médicale de Bergerac* (Paris: Vigot, 1936).

49. Maine de Biran, "Observations sur les divisions organiques du cerveau considérées comme sièges des différentes facultés intellectuelles et morales," in *Tome V des œuvres de Maine de Biran: Discours à la société médicale de Bergerac*, ed. François Azouvi (Paris: J. Vrin, 1984), 76.

50. See Robert Young, *Mind, Brain, and Adaptation in the Nineteenth Century* (Oxford: Clarendon, 1970), 9–53.

51. The downfall of cranioscopy as a pseudoscience can be attributed to a biting

satire published by John Gordon, "The Doctrines of Gall and Spurzheim," *Edinburgh Review* 25 (1815): 227–68.

52. See F. J. Gall and G. Spurzheim, *Recherches sur le système nerveux en général, et sur celui du cerveau en particulier: Présenté à l'Institut de France, le 14 mars 1808* (Paris: Schoell, 1809); Biran transmitted Gall's studies to Bergerac by way of Georges Cuvier's review, *Rapport sur un mémoire de MM. Gall et Spurzheim, relatif à l'anatomie du cerveau* (Paris: Baudouin, 1808).

53. The French alienists Jules Baillarger (1809–90) and Jacques-Joseph Moreau de Tours (1804–84) confirmed the medical value of Biran's division between the intellectual and affective faculties. These clinicians examined cases of mental regression, which they believed to reveal the involuntary dimension of consciousness. Baillarger was known for his studies of hallucinations, depression, and what is now called manic-depressive disorder. These delirious states evinced what he took to be psychological automatism. In *Du hachisch et de l'aliénation mentale: Etudes psychologiques* (Paris: Fortin Massin, 1845), Moreau de Tours suggested that the drug offered a glimpse into affective experience by dampening the motor activity of attentive consciousness. He drew on Biran's division, arguing that affection resurfaces in the absence of intellection. Baillarger and Moreau de Tours showed that mental maladies constitute a positive expression of psychic activity and not a failure of attention. See Henri Delacroix, "Maine de Biran et l'école médico-psychologique," *Bulletin de la Société française de philosophie* 24, nos. 2–3 (1924): 51–63.

54. Biran, "Observations sur les divisions organiques du cerveau," 74.

55. Ibid., 80.

56. See Daniel N. Robinson, *An Intellectual History of Psychology*, 3rd ed. (Madison: University of Wisconsin Press, 1995), 149–96.

57. Maine de Biran, *Nouvelles considérations sur les rapports du physique et du moral de l'homme*, ed. Victor Cousin (Paris: Ladrange, 1834). The patrimony of Biran's writings is documented in the introduction of this volume. Also see the subsequent volume, Maine de Biran, "Nouvelles considérations sur le sommeil et les songes," in *Mémoires de l'académie royale des sciences morales et politique*, vol. 2, ed. Victor Cousin (Paris: Didot, 1837).

58. Victor Cousin, introduction to *Œuvres philosophiques de Maine de Biran*, vol. 4 (Paris: Ladrange, 1841), x.

59. Pierre Leroux, "De la mutilation d'un écrit posthume de M. Jouffroy," *Revue indépendante*, November 1, 1842, 293. I am indebted to Warren Breckman for my knowledge of Leroux. See Breckman's "Politics in a Symbolic Key: Pierre Leroux, Romantic Socialism, and the Schelling Affair," *Modern Intellectual History* 2, no. 1 (2005): 61–86.

60. Patrice Vermeren, *Victor Cousin: Le jeu de la philosophie et de l'état* (Paris: L'Harmattan, 1995), 208.

61. Janet, *Les problèmes du XIXe siècle*, 285.

62. Maine de Biran, *Œuvres philosophiques de Maine de Biran*, ed. Victor Cousin (Paris: Ladrange, 1841).

63. Ibid., xxvii.

64. There were occasions when Cousin did formulate his eclectic spiritualism in terms that were faithful to Biran, such as when he wrote: "The will alone is the person or the *moi*." Cousin, "Préface de la deuxième édition," *Fragments philosophiques*, xxxv.

65. The divergence between Biran's and Cousin's accounts of self-consciousness

also turned on their respective readings of Leibniz. In *La Monadologie* (1714), Leibniz claimed that consciousness entails an act of apperception. Consciousness is *of* an object. Analogously, Biran conceptualized selfhood as a motor force that encounters the phenomenon of resistance. Cousin, however, conceived the self as a substance, a foundation he believed necessary to secure the identity of personhood. Pierre Leroux was critical that Cousin thus reified consciousness: "only such a hardly solid thinker could . . . take consciousness or apperception for something isolable from the phenomenon." Pierre Leroux, *Réfutation de l'éclecticisme* (Paris: Charles Gosselin, 1841), 129.

66. Cousin, "Préface à la premier édition," in *Fragments philosophiques*, 39.

67. John Locke, *Essai philosophique concernant l'entendement humain*, trans. Pierre Coste (Amsterdam: P. Mortier, 1714), 265. Also see John Locke, *Identité et différence: L'invention de la conscience*, ed. Étienne Balibar (Paris: Éditions du Seuil, 1998).

68. See Sophia A. Rosenfeld, *Common Sense: A Political History* (Cambridge, MA: Harvard University Press, 2011); James W. Manns, *Reid and His French Disciples* (Leiden: E. J. Brill, 1994); Manns, *Victor Cousin, les idéologues et les écossais: Colloque international* (Paris: Presses de l'École normale supérieure, 1985); Edward H. Madden, "Victor Cousin and the Commonsense Tradition," *History of Philosophy Quarterly* 1, no. 1 (1984): 93–109.

69. Victor Cousin, *Elements of Psychology*, trans. Caleb S. Henry (New York: Ivison and Phinney, 1862), 406.

70. Following her travels to Germany in 1803 and 1807, de Staël compiled her reflections on transcendental philosophy in *De l'allemagne* (Paris: H. Nicolle, 1810), a book censored for allegedly adding fuel to the fire of France's perpetual conflicts beyond the Rhine.

71. Charles Adam, *La philosophie en France* (Paris: Félix Alcan, 1894), 22–23.

72. See Victor Cousin, "Souvenirs d'Allemagne," *Revue des deux mondes*, no. 64 (1866): 594–620.

73. Paul Janet, *Victor Cousin et son oeuvre* (Paris: Calmann-Lévy, 1885), 82.

74. Elme-Marie Caro, "De la situation actuelle du spiritualisme," *Revue des cours littéraires de la France et de l'étranger*, December 24, 1864, 58.

75. Ibid. Caro systematized his remarks on spiritualism in his treatise *Le matérialisme et la science* (Paris: Hachette, 1867).

76. Ernest Naville, *Notice historique et bibliographique sur les travaux de Maine de Biran* (Paris: Avril, 1851), vii.

77. François Naville, ed., "Fragments inédits de Maine de Biran," *Bibliothèque universelle de Genève*, vols. 61–63 (Paris: Chez Anselin, 1845).

78. Ernest Naville published Biran's *Journal intime* (Paris: J. Cherbuliez, 1857). Also see Ernest Naville, "Histoire des manuscrits inédits de Maine de Biran," in *Maine de Biran, sa vie et ses pensées*, xxix.

79. Chanoine Mayjonade, "L'évolution religieuse de Maine de Biran," *Bulletin de la Société française de philosophie* 24, nos. 2–3 (1924): 66.

80. Maine de Biran, *Science et psychologie: Nouvelle œuvres inédites de Maine de Biran*, ed. Alexis Bertrand (Paris: Ernest Leroux, 1887).

81. See Alexis Bertrand, "Introduction," in ibid., i–xxxiv.

82. Seigel, *Idea of the Self*, 251.

83. Joseph Ferrari, *Les philosophes salariés*, 1849, ed. Stéphane Douailler and Patrice Vermeren (Paris: Payot, 1983), 88. Ferrari was a socialist philosophy professor

at the Université de Strasbourg. He was ejected from his post for having taught a controversial course on Renaissance thought.

84. Ravaisson translated and published Schelling's brief introduction to Cousin's work, "Jugement de Schelling sur la philosophie de M. Cousin, et sur l'état de la philosophie française et de la philosophie allemande en général," in *Revue germanique* 3, no. 10 (1835): 3–24.

85. For Schelling, "art is paramount to the philosopher, precisely because it opens up to him, as it were, the holy of holies, where burns in eternal and original unity, as if in a single flame, that which in nature and history is rent asunder, and in life and action, no less than in thought, must forever fly apart." *System of Transcendental Idealism*, 1800, trans. Peter Heath (Charlottesville: University Press of Virginia, 1978), 232.

86. Félix Ravaisson, "Fragments de philosophie de M. Hamilton," *Revue des deux mondes* 24 (1840): 422.

87. Ibid., 425.

88. Jules Lachelier, letter to Paul Janet, December 8, 1887, MS 4687, Opuscules et correspondance de Jules Lachelier, Bibliothèque de l'Institut de France.

89. Félix Ravaisson, *Of Habit*, trans. Mark Sinclair and Claire Carlisle (London: Continuum, 2009), 59; originally published as *De l'habitude* (Paris: Fournier, 1838).

90. Mark Sinclair, "Ravaisson and the Force of Habit," *Journal of the History of Philosophy* 49, no. 1 (2011): 75.

91. Ravaisson overcame the modesty of the vitalists that he cited. These thinkers posited the vital principle as one element within nature, as if vitality occupied an island in a sea of physical matter. Instead, Ravaisson treated the totality of nature as the manifestation of a single vital principle that progressively enlarged its scope across the Aristotelian strata of vegetal, animal, and human life. See Annie Bitbol-Hespériès, "Ravaisson et la philosophie de la médecine," in *L'épistémologie française, 1830–1970*, ed. Michael Bitbol and Jean Gayon (Paris: Presses universitaires de France, 2006), 413–30; Jean Cazeneuve, *La philosophie médicale de Ravaisson* (Paris: Presses universitaires de France, 1958).

92. Ravaisson, *Of Habit*, 31.

93. Ibid., 29.

94. Ibid., 59.

95. Claire Marin, "Ravaisson et Bergson: La Science du Vivant," in *Annales bergsoniennes III: Bergson et la science*, ed. Frédéric Worms (Paris: Presses universitaires de France, 2007), 387.

96. Ravaisson, *Of Habit*, 59.

97. Pierre Montebello, *L'autre métaphysique: Essai sur la philosophie de la nature; Ravaisson, Tarde, Nietzsche et Bergson* (Paris: Desclée de Brouwer, 2003), 14.

98. Ravaisson, *Of Habit*, 45.

99. Ibid., 57.

100. For an inventive reading of the aesthetics of grace in Ravaisson's thought, see Michael Fried, *Flaubert's "Gueuloir": On Madame Bovary and Salammbô* (New Haven, CT: Yale University Press, 2012); *Courbet's Realism* (Chicago: University of Chicago Press, 1992), 182–87.

101. Ravaisson, *Of Habit*, 71.

102. Félix Ravaisson, *L'enseignement du dessin dans les lycées* (Paris: Ministère de l'instruction publique et des cultes, 1854), 18.

103. Ravaisson composed a drawing textbook, which included images of classical

sculptures as well as modern sketches. Students were expected to draw from these examples in the classroom. Félix Ravaisson, *Les classiques de l'art: Modèles pour l'enseignement du dessin* (Paris: Bibliothèque nationale de France, 1853). He elaborated on his philosophy of aesthetics as a curator of classical antiques at the Louvre, a position he took over in 1870. See Félix Ravaisson, *La Vénus de Milo* (Paris: Hachette, 1871). A valuable exposition of his aesthetics is in Anne Henry, *Marcel Proust: Théories pour une esthétique* (Paris: Klincksieck, 1983), 81–97, 131–35.

104. Ravaisson, *L'enseignement du dessin dans les lycées*, 66.

105. Bergson, "Life and Work of Ravaisson," 197.

106. See Vermeren, *Victor Cousin*, 217.

107. See Mark Sinclair, "Editor's Introduction," in *Of Habit*, trans. Mark Sinclair (New York: Continuum, 2008), 1–21.

108. Jules Lachelier, letter to Paul Janet, December 8, 1887, MS 4687, Opuscules et correspondance de Jules Lachelier, Bibliothèque de l'Institut de France.

109. Lachelier, *Du fondement de l'induction*, 102, my emphasis.

110. Jules Lachelier, letter to Félix Ravaisson, December 5, 1859, MS 4687, Opuscules et correspondance de Jules Lachelier, Bibliothèque de l'Institut de France.

111. For a contextual reconstruction of Biran's account of causation, see Philip P. Hallie, "Hume, Biran, and the *Méditatifs intérieurs*," *Journal of the History of Ideas* 18, no. 3 (1957): 295–312.

112. Jules Lachelier, letter to Félix Ravaisson, December 5, 1859, MS 4687, Opuscules et correspondance de Jules Lachelier, Bibliothèque de l'Institut de France.

113. "Université de Paris: Agrégés de 1863," 61 AJ 47, Archives nationales, Paris.

114. Jules Lachelier, letter to Félix Ravaisson, August 15, 1868, MS 4687, Opuscules et correspondance de Jules Lachelier, Bibliothèque de l'Institut de France.

115. Célestin Bouglé, *Les maîtres de la philosophie universitaire en France* (Paris: Librairie Maloine, 1938), 5.

116. Immanuel Kant, *Critique of Pure Reason*, trans. Norman Kemp Smith (New York: Palgrave Macmillan, 1929), A91/B124. Philosophers of science in the twentieth century largely abandoned these conceptual requirements. Karl Popper demonstrated that scientists do not rely on induction. Instead, they invoke verification and falsification. Popper contended that the sciences arrive at theories whose predictive value is the subject of relentless testing. See Popper, *The Logic of Scientific Discovery*, 1934 (New York: Routledge, 2002).

117. Claude Bernard, *An Introduction to the Study of Experimental Medicine*, 1865. trans. Henry Copley Greene (New York: Macmillan, 1927), 93.

118. Lachelier, *Du fondement de l'induction*, 73.

119. Kant, *Critique of Pure Reason*, A91/B124.

120. Lachelier, *Du fondement de l'induction*, 35.

121. Louis Milet, *Le symbolisme dans la philosophie de Lachelier* (Paris: Presses universitaires de France, 1959), 12.

122. Lachelier, *Du fondement de l'induction*, 51.

123. Ibid., 56.

124. Ibid., 81.

125. The necessary determination of natural phenomena had originated in the Enlightenment and continued well into the nineteenth century. See Lorraine Daston and Cianna Pomata, "The Faces of Nature: Visibility and Authority," in *The Faces of Nature in Enlightenment Europe*, ed. Daston and Pomata (Berlin: BWV, 2003), 1–16.

126. Alphonse Darlu, "Réflexions d'un philosophe sur les questions du jour: Science, morale et religion," *Revue de métaphysique et de morale* 3 (1895): 249; also see H. W. Paul, "The Debate over the Bankruptcy of Science," *French Historical Studies* 5 (1968): 299–337.

127. Jules Lachelier, letter to Félix Ravaisson, December 21, 1867, MS 4687, Opuscules et correspondance de Jules Lachelier, Bibliothèque de l'Institut de France.

128. Jules Lachelier, letter to Gabriel Séailles, October 15, 1913, MS 4687, Opuscules et correspondance de Jules Lachelier, Bibliothèque de l'Institut de France.

129. Lachelier, *Du fondement de l'induction*, 86.

130. On Boutroux's Catholicism, see Joel Revill, "Émile Boutroux, Redefining Science and Faith in the Third Republic," *Modern Intellectual History* 6, no. 3 (2009): 485–512.

131. Émile Boutroux, "Notes autobiographique, 1909–1912," MS 4122.5, Cahiers d'étudiant et papiers divers d'Émile Boutroux, Bibliothèque de l'Institut de France.

132. Émile Boutroux, letter to Jules Lachelier, 1868, MS 4122.7, Cahiers d'étudiant et papiers divers d'Émile Boutroux, Bibliothèque de l'Institut de France.

133. Ibid.

134. Jules Lachelier, letter to Émile Boutroux, 1868, MS 4687, Opuscules et correspondance de Jules Lachelier, Bibliothèque de l'Institut de France.

135. Jules Lachelier, letter to Émile Boutroux, September 6, 1868, MS 4687, Cahiers d'étudiant et papiers divers d'Émile Boutroux, Bibliothèque de l'Institut de France.

136. Émile Boutroux, "La vie universitaire en Allemagne," *Revue politique et littéraire* 1 (1871): 542–49.

137. Boutroux dedicated his thesis to Ravaisson. He also closely studied *Of Habit*. In his personal copy of the book, Boutroux emphatically underlined the double law of habit: "The continuity or repetition of the passion weakens it: the continuity or repetition of the action exalts and strengthens it." For Ravaisson, the double law reflected his method of tracing the conditions of human freedom from the animating energy of nature. In his lone marginal note, Boutroux wrote: "what precisely could explain the difference between habit and the will?" The note suggests that the young philosopher was preoccupied by the corporeal account of selfhood. Émile Boutroux, "Transcription of *De l'habitude*," 15, MS 4122.1, Cahiers d'étudiant et papiers divers d'Émile Boutroux, Bibliothèque de l'Institut de France.

138. Auguste Comte, *Cours de philosophie positive*, 6 vols. (Paris: Bachelier, 1830–42).

139. See Simon, *European Positivism in the Nineteenth Century*.

140. Boutroux, *Contingency of the Laws of Nature*, 158–59.

141. Ibid., 30.

142. Ibid. Boutroux's argument that the nature of the conditioned is to exceed its conditions prefigured a key ontological commitment of French poststructuralist thought. It was central to Gilles Deleuze's thesis that difference precedes identity, as when he claimed that difference "affirms at once both the unconditioned character of the product in relation to the conditions of production, and the independence of the work in relation to its author or actor" (*Difference and Repetition*, 1968, trans. Paul Patton [New York: Continuum, 2007], 94). Although I am unaware of whether he had read *The Contingency of the Laws of Nature*, Deleuze could have enriched his argument had he broached a conversation with Boutroux's concept of heterogeneity.

143. Boutroux, *Contingency of the Laws of Nature*, 69–70.

144. Ibid., 66.

145. Ibid., 160.

146. Ibid., 171.

147. Émile Boutroux, "Rapport de la pensée et des choses dans l'idéalisme," MS 4120.2, Cahiers d'étudiant et papiers divers d'Émile Boutroux, Bibliothèque de l'Institut de France.

148. Boutroux, *Contingency of the Laws of Nature*, 172.

149. Émile Boutroux, "Jules Lachelier," *Revue de métaphysique et de morale* 28, no. 1 (1921): 6.

150. Jules Lachelier, "Spiritualism," in *Vocabulaire technique et critique de la philosophie*, ed. André Lalande (Paris: Félix Alcan, 1926), 792–93.

151. Ravaisson, *Rapport sur la philosophie en France au XIXe siècle*, 283.

Chapter Two

1. Théodule Ribot, "De la durée des actes psychiques d'après des travaux récents," *Revue philosophique* 1 (1876): 274.

2. Jules Tannery, "Correspondance: A propos du logarithme des sensations," *Revue scientifique* 4, no. 37 (1875): 16.

3. Émile Boutroux, *Natural Law in Science and Philosophy*, trans. Fred Rothwell (New York: Macmillan, 1914), 183; originally published as *De l'idée de loi naturelle dans la science et la philosophie contemporaines* (Paris: Félix Alcan, 1895).

4. Henri Bergson, *Time and Free Will: An Essay on the Immediate Data of Consciousness*, 3rd ed., trans. F. L. Pogson (London: George Allen, 1913); originally published as *Essai sur les donnés immédiates de la conscience* (Paris: Félix Alcan, 1889).

5. See Philippe Soulez and Frédéric Worms, *Bergson* (Paris: Presses universitaires de France, 2002).

6. Jacques Chevalier, *Entretiens avec Bergson* (Paris: Plon, 1959), 19. Joseph Desaymard, a student of Bergson, recounts the same transformation in "H. Bergson à Clermont-Ferrand," in *Bulletin historique et scientifique d'Auvergne* (Clermont-Ferrand: Bellet, 1910): 216.

7. Alfred Fouillée, *La psychologie des idées-forces*, vol. 2 (Paris: Félix Alcan, 1893), 29.

8. Jules Lachelier, *Psychologie et métaphysique* (Paris: Presses universitaires de France, 1949), 4; originally published in *Revue philosophique* 19 (1885): 481–516.

9. See Wilhelm Wundt, *Über den Einfluss der Philosophie auf die Erfahrungswissenschaft* (Leipzig: Engelmann, 1876).

10. Lachelier, *Psychologie et métaphysique*, 71.

11. Serge Nicolas and Rasyid Bo Saitioso, "Alfred Binet and Experimental Psychology at the Sorbonne Laboratory," *History of Psychology* 15, no. 4 (2012): 328–63.

12. The Sorbonne laboratory was the thirteenth in the world founded after Wundt's 1879 laboratory. See William S. Sahakian, *History and Systems of Psychology* (New York: Wiley, 1975), 138–39.

13. The origin story of 1879 can be traced to an article by Wundt's student James McKeen Cattell, "The Psychological Laboratory at Leipsic [*sic*]," *Mind* 13 (1888): 37–51.

14. Serge Nicolas claims that spiritualist thinkers delayed the reception of experi-

mental psychology in France. See Nicolas, *Les facultés de l'âme: Une histoire des systèmes* (Paris: L'Harmattan, 2005); Nicolas, "Introducing Psychology as an Academic Discipline in France: Théodule Ribot and the Collège de France (1881–1901)," *Journal of the History of the Behavioral Sciences* 37, no. 2 (2001): 143–64. The view is also discussed in Jacqueline Carroy and Régine Plas, "The Origins of French Experimental Psychology: Experiment and Experimentalism," *History of the Human Sciences* 9, no. 1 (1996): 73–84.

15. This chapter builds on the work of historians who have debunked the patricidal narrative according to which experimental psychology liberated itself from metaphysical psychology. See Dorothy Ross, ed., *Modernist Impulses in the Human Sciences, 1870–1930* (Baltimore: Johns Hopkins University Press, 1994); John L. Brooks III, *The Eclectic Legacy: Academic Philosophy and the Human Sciences in Nineteenth-Century France* (Newark: University of Delaware Press, 1998); Edward S. Reed, *From Soul to Mind: The Emergence of Psychology from Erasmus Darwin to William James* (New Haven, CT: Yale University Press, 1997).

16. Théodule Ribot, *L'hérédité, etude psychologique* (Paris: Ladrange, 1873), 302.

17. Ibid., 353.

18. I have in mind the work of Franz Brentano, Wilhelm Dilthey, Carl Stump, and Edmund Husserl.

19. In Anglo-American philosophy, the "hard problem" of consciousness can be traced back to the seminal article by Thomas Nagel, "What Is It Like to Be a Bat?," *Philosophical Review* 83, no. 4 (1974): 435–50.

20. Henri de Parville, "Physiologie et médecine," *Causeries scientifiques* 9 (Paris: J. Rothschild, 1869): 187.

21. Johannes Peter Müller, *Handbuch der Physiologie des Menschen für Vorlesungen*, 2 vols. (Coblenz: J. Hölscher, 1837–40).

22. Later research revealed that nerves carry the same action potentials; they differ in the brain, which coordinates the distinct sensory qualities. See Edgar Douglas Adrian, *The Basis of Sensation: The Action of the Sense Organs* (London: Christophers, 1928).

23. See Nima Bassiri, "Material Translations in the Cartesian Brain," *Studies in History and Philosophy of Biological and Biomedical Sciences* 43 (2012): 244–55.

24. Carlo Matteucci, *Lezioni di fisica* (Pisa: Presso Rocco Vannucchi, 1841).

25. See Gabriel Finkelstein, *Emil du Bois-Reymond: Neuroscience, Self, and Society in Nineteenth-Century Germany* (Cambridge, MA: MIT Press, 2013).

26. Du Bois-Reymond compared the speed of nerve propagation with the speed of electricity and physical objects. See "Vitesse de la transmission de la volonté et de la sensation à travers les nerfs: Conférence de M. du Bois-Reymond à l'Institution royale de la Grande-Bretagne," *Revue scientifique* 4, no. 3 (1866): 33–41.

27. The term "physiological time" developed out of the "personal equation" in astronomy. The latter referred to the inconsistencies in astronomical measurements among different observers. Their assessments of stars' and planets' movements resulted in variations in the portions of seconds obtained; however, there was little variation in the same person's assessments. As a result, psychologists set about measuring the tenths of a second it took the body to carry out such observations. See Christoph Hoffmann, "Constant Differences: Friedrich Wilhelm Bessel, the Concept of the Observer in Early Nineteenth-Century Practical Astronomy and the History

of the Personal Equation," *British Journal for the History of Science* 40, no. 3 (2007): 333–65; Simon Schaffer, "Astronomers Mark Time: Discipline and the Personal Equation," *Science in Context* 2, no. 1 (1988): 115–45.

28. Adolph Hirsch, "Sur les corrections et équations personnelles dans les observations chronographiques de passage," *Bulletin de la Société des sciences naturelles de Neuchâtel* 6 (1864): 300.

29. Hermann von Helmholtz, "Vorläufiger Bericht über die Fortpflanzungsgeschwindigkeit der Nervenreizung," *Archiv für Anatomie, Physiologie und wissenschaftliche Medicin* 20 (1850): 71–73. According to Henning Schmidgen, Helmholtz brought about a new mode of representation. He did not make use of surveys that were believed to accurately depict the world to measurement but instead used graphic methods where "what was measured was produced by the measurement." See Schmidgen, *The Helmholtz Curves: Tracing Lost Time*, trans. Nils F. Schott (New York: Fordham University Press, 2014).

30. Afterward the Viennese scientist Sigmund Exner invented the term "reaction time." His own experiments measured the involuntary reaction time to electric stimuli. See Exner, "Experimentelle Untersuchung der einfachsten psychischen Prozesse: Erste Abhandlung; Die persönliche Gleichung," *Archiv für die gesamte Physiologie des Menschen und der Thiere herausgegeben* 7 (1873): 608–9.

31. Adolph Hirsch, "Expériences chronométriques sur la vitesse des différentes sensations et de la transmission, nerveuse," *Bulletin de la Société des sciences naturelles de Neuchâtel* 6 (1862): 100–114.

32. See Jimena Canales, *A Tenth of a Second: A History* (Chicago: University of Chicago Press, 2009), especially 29–58.

33. Etienne-Jules Marey, "Leçon d'ouverture: Vitesse des actes nerveux et cérébraux—Le vol dans la série animale, Collège de France, histoire naturelle des corps organisés, cours de M. Marey," *Revue scientifique* 6, no. 4 (1868): 61–64.

34. Hermann von Helmholtz, "Über die Zeit, welche nötig ist, damit ein Gesichtseindruck zum Bewusstsein kommt, Resultate einer von Herrn N. Baxt aus Petersburg im Heidelberger Laboratorium ausgegührten Untersuchung," in Hermann von Helmholtz, ed., *Wissenschaftliche Abhandlungen*, vol. 2 (Leipzig: J. A. Barth, 1883), 947–52.

35. F. C. Donders, "On the Speed of Mental Processes," trans. W. G. Koster, *Acta psychologica* 30 (1969): 413; originally published as "Over de snelheid van psychische processen," *Nederlandsch Archief voor Genees- en Natuurkunde* 4 (1869): 117–45. Also see Henning Schmidgen, "The Donders Machine: Matter, Signs and Time in a Physiological Experiment, ca. 1865," *Configurations* 13, no. 2 (2005): 2–56.

36. F. C. Donders, "On the Speed of Mental Processes."

37. Rodolphe Radau, "La vitesse de la volonté," in *Les derniers progrès de la science* (Paris: Leiber, 1868), 221.

38. Robert M. Young confirms, "The significance of the nineteenth-century analysis [of the sensory-motor framework] lay first in its experimental demonstration in the central nervous system and second in the progressive extension of the concept as the fundamental explanatory principle in both physiology and psychology." Young, *Mind, Brain, and Adaptation*, 93.

39. Albert René, "Étude expérimentale sur la vitesse de transmission nerveuse chez l'homme: Durée d'un acte cérébral et d'un acte réflexe, vitesse sensitive, vitesse motrice," *Gazette des hôpitaux* 55 (1882): 276–77.

40. Ludwig Lange, "Neue Experimente über den Vorgang der einfachen Reaction auf Sinneseindrücke," *Philosophische Studien* 4 (1888): 479–510.

41. Reaction experiments were only part of the Leipzig research program. See Arthur L. Blumenthal, "A Reappraisal of Wilhelm Wundt," *American Psychologist* 30 (1975): 1081–88.

42. See Serge Nicolas and Ludovic Ferrand, "Wundt's Laboratory at Leipzig in 1891," *History of Psychology* 2, no. 3 (1999): 194–203.

43. See Serge Nicolas and Peter B. Thompson, "The Hipp Chronoscope versus the D'Arsonval Chronometer. Laboratory Instruments Measuring Reaction Times That Distinguish German and French Orientations of Psychology," *History of Psychology* 18, no. 4 (2015): 367–84.

44. See Jacqueline Carroy and Henning Schmidgen, "Reaction Time Tests in Leipzig, Paris and Würzburg: The Franco-German History of a Psychological Experiment, 1890–1910," *Medizinhistorisches Journal* 39, no. 1 (2004): 27–55.

45. E. Bradford Titchener, "A Psychological Laboratory," *Mind* 7 (1898): 311–31.

46. See Henri Beaunis, "Sur le temps de réaction des sensations olfactives," *Comptes rendus des séances de l'Académie des sciences* 96 (1883); Léon Lalanne, "Note sur la durée de la sensation tactile," *Journal de l'anatomie et de la physiologie normales et pathologiques de l'homme et des animaux* (Paris: Charles Robin, 1876): 448–57.

47. The *Review of Experimental Psychology* (*Revue de psychologie expérimentale*) had been founded in 1874 by the French physician Timothée Puel (1812–90). The journal's subtitle was "Studies on Sleep, Somnambulism, Hypnotism and Spiritualism." The reference to spiritualism should not be confused with the philosophical movement. Puel brought together articles on the *spiritism* of Hippolyte Léon Denizard Rivail (1804–69). Under the pseudonym Allan Kardec, he popularized the study of immortal spirits. Puel was interested in the use of mediums and séances to investigate occult phenomena. He also integrated experimental studies on mental pathologies. But the journal lacked the support of the academic establishment and, as a result, lasted merely two years. See J. W. Monroe, *Laboratories of Faith*.

48. Notable examples of French psychometric reports include Alfred Binet, "La perception de la durée dans les réactions simples," *Revue philosophique de la France et de l'étranger* 33 (1892): 650–59; Henri Beaunis, "Influence de la durée de l'expectation sur le temps de réaction des sensations visuelles," *Revue philosophique* 20 (1885): 330–32; Beaunis, "Sur la comparaison du temps de réaction des différentes sensations," *Revue philosophique* 15 (1883): 611–20; Charles Richet, "De la durée des actes psychiques élémentaires," *Revue philosophique* 6 (1878): 393–96; Léon Lalanne, "Note sur la durée de la sensation tactile," *Revue philosophique* 2 (1876): 650.

49. Théodule Ribot, "Introduction," *Revue philosophique* 1 (1876): 3.

50. Ibid.

51. Ibid., 2.

52. Théodule Ribot, letter to Alfred Espinas, April 1876, "Lettres de Théodule Ribot à Alfred Espinas (1876–1893)," ed. Raymond Lenoir, *Revue philosophique* 152 (1962): 338.

53. Théodule Ribot, *English Psychology* (New York: D. Appleton, 1874), 25 (translation modified); originally published as *La psychologie anglaise contemporaine* (Paris: Ladrange, 1870).

54. See Jacqueline Thirard, "La fondation de la 'Revue philosophique,'" *Revue philosophique* 160 (1976): 401–13.

55. Théodule Ribot, letter to Charles Lévêque, August 21, 1874, MS 2563, Papiers de Charles Lévêque, Bibliothèque de l'Institut de France, Paris.

56. Ribot, "De la durée des actes psychiques," 268.

57. Théodule Ribot, *La psychologie allemande contemporaine (école expérimentale)* (Paris: Librairie Germer Baillière et Cie, 1879). Critics alleged that Ribot's narrow focus on reaction experiments had excluded the wide range of Wundt's interests, especially his work in aesthetics, religion, and *Völkerpsychologie*. Wundt took note. In the preface to the French translation of his psychology handbook, Wundt emphasized that his objective was to understand laws specific to the mind, apart from their neurological underpinnings. He was critical of psychologists who "imagine that physiological psychology has the intention of replacing psychology with a physiology of the brain." Wilhelm Wundt, "Préface de l'auteur pour l'édition française," *Éléments de psychologie physiologique* (Paris: Félix Alcan, 1886), xxvii.

58. Ernst Heinrich Weber, *De pulsu, resorptione, auditu et tactu: Annotationes anatomicae et physiologicae* (Leipzig: C. F. Koehler, 1834).

59. Jean-Baptiste Biot, *Lehrbuch der Experimental-Physik oder Erfahrungs-Naturlehre*, vol. 3 (Leipzig: Leopold Voß, 1829). Although the five-volume series was published under Biot's name, Fechner wrote the entire third volume.

60. Gustav Fechner, *Elements of Psychophysics*, trans. Helmut E. Adler, ed. Davis. H. Howes and Edwin G. Boring (New York: Holt, Rinehart and Winston, 1966), 6; originally published as *Elemente der Psychophysik: Erster und zweiter Thiel* (Leipzig: Breitkopf und Härtel, 1860).

61. Fechner, *Elements of Psychophysics*, 7.

62. See Joel Michell, *Measurement in Psychology: A Critical History of a Methodological Concept* (Cambridge: Cambridge University Press, 1999), 81.

63. For a thorough mathematical explication of Fechner's law, see Michael Heidelberger, *Nature from Within: Gustav Theodor Fechner and His Psychophysical Worldview*, trans. Cynthia Klohr (Pittsburgh: University of Pittsburgh Press, 2004), 200–207.

64. Before Einstein's relativity thesis, nineteenth-century scientists took time to be homogeneous throughout (at least within the limited range of the earth's temporality). It could be broken up into equal parts; and those parts could be measured using the uniform distance between the points on the clock.

65. See Sergio Cesare Masin, Verena Zudini, and Mauro Antonelli, "Early Alternative Derivations of Fechner's Law," *Journal of the History of the Behavioral Sciences* 45, no. 1 (2009): 56–65.

66. Joseph Delboeuf, "Étude psychophysique: Recherches théorique et expérimentales sur la mesure des sensations et spécialement des sensations de lumière et de fatigue," in *Mémoires couronnées et autres mémoires publiés par l'Académie royale de Belgique*, vol. 26 (Brussels: Hayez, 1873); Joseph Plateau, "Sur la mesure des sensations physiques et sur la loi qui lie l'intensité de ces sensations à l'intensité de la cause excitante," *Bulletin de l'Académie royale de Belgique* 33 (1872): 376–88.

67. Joseph Delboeuf, *Études psychophysiques sur la mesure des sensations* (Brussels: Hayez, 1873).

68. Joseph Delboeuf, *Éléments de psychophysique* (Paris: Baillière, 1883).

69. Joseph Delboeuf, "La loi psychophysique: Hering contre Fechner," *Revue philosophique* 3 (1877): 225–63; Delboeuf, "La loi psychophysique et le nouveau livre de Fechner—I," *Revue philosophique* 5 (1878): 34–63; Delboeuf, "La loi psychophysique

et le nouveau livre de Fechner—II," *Revue philosophique* 5 (1878): 127–57; Delboeuf, "Le sentiment de l'effort," *Revue philosophique* 12 (1881): 513–27.

70. Ribot, "De la durée des actes psychiques," 274.

71. Ibid., 287.

72. Jules Tannery, "Correspondance: A propos du logarithme des sensations," *Revue scientifique* 4, no. 37 (1875): 16–17.

73. In Germany, Johannes Von Kries (1853–1928) similarly argued that extensive quantities such as length, mass, and time can be measured but intensive quantities cannot. See Johannes Von Kries, "Über die Messung intensiver Grössen und über das sogenannte psychophysische Gesetz," *Vierteljahrsschrift für wissenschaftliche Philosophie* 6 (1882): 257–94.

74. Tannery, "Correspondance," 17.

75. Paul Tannery, "Critique de la loi de Weber," *Revue philosophique* 17 (1884): 15–35. Tannery spent little time in academia. He completed an engineering degree at the École polytechnique and briefly taught ancient history at the Sorbonne before leaving to pursue a career in the state-run tobacco industry.

76. Jules Tannery, letter to Paul Tannery, 1874, in Paul Tannery, *Mémoires scientifiques*, vol. 15 (Paris: Cauthier-Villars, 1939), 94.

77. Émile Bordel, "Introduction," in Jules Tannery, *Science et philosophie*, ed. Émile Bordel (Paris: Félix Alcan, 1912), xiii.

78. Jules Tannery, "Correspondance," 17.

79. Théodule Ribot, "Correspondance: A propos du logarithme des sensations," *Revue scientifique* 4, no. 37 (1875): 17–18.

80. Joseph Delboeuf, "Correspondance: A propos du logarithme des sensations," *Revue scientifique* 4, no. 43 (1875): 11.

81. Wilhelm Wundt, "Correspondance: A propos du logarithme des sensations," *Revue scientifique* 4, no. 43 (1875): 13–14.

82. Jules Tannery, "Correspondance: A propos du logarithme des sensations," *Revue scientifique* 4, no. 43 (1875): 15.

83. For a thorough discussion of Fechner's aesthetics, see Fernando Vidal, "La neuroésthetique, un esthétisme scientiste," *Revue d'histoire des sciences humaines* 2, no. 25 (2011): 239–64.

84. Émile Boutroux, *Jules Tannery 1848–1910*, manuscript found in Manuscrits Émile Boutroux, École normale supérieure, Paris.

85. Mary Joe Nye documented the "Boutroux circle" comprising Boutroux, the Tannery brothers, and the mathematicians Benjamin Baillaud and Henri Poincaré. The circle was committed to conventionalism, the idea that the veracity of scientific models was not a matter of their correspondence with nature, but whether they offered conventional tools to explain natural processes. See Mary Jo Nye, "The Boutroux Circle and Poincaré's Conventionalism," *Journal of the History of Ideas* 40, no. 1 (1979): 107–20.

86. Émile Boutroux, letter to Paul Tannery, April 26, 1873, MS 4122.2, Cahiers d'étudiant et papiers divers d'Émile Boutroux, Bibliothèque de l'Institut de France.

87. Ibid.

88. Ibid.

89. Ibid.

90. Jules Tannery, letter to Émile Boutroux, April 1873, MS 4122.2, Cahiers d'étudiant et papiers divers d'Émile Boutroux, Bibliothèque de l'Institut de France.

91. Émile Boutroux, *Natural Law in Science and Philosophy*, trans. Fred Rothwell (New York: Macmillan, 1914), 184. The book comprises Boutroux's lectures delivered at the Sorbonne in 1892–93.

92. Martin Trautscholdt, "Experimentelle Untersuchungen über die Association der Vorstellungen," *Philosophische Studien* 1 (1883): 213–50.

93. Charles Féré, *Sensation et mouvement: Études expérimentales du psychomécanique* (Paris: Félix Alcan, 1887).

94. Alexis Bertrand, *La psychologie de l'effort et les doctrines contemporaines* (Paris: Félix Alcan, 1889), 104.

95. A.-M. Bloch, "Expériences sur la vitesse du courant nerveux sensitif de l'homme," *Archives de physiologie normale et pathologique, publiées par MM. Brown-Séquard, Charcot, Vulpian* 2 (1875): 622.

96. A.-M. Bloch, "Psychologie: La vitesse comparative des sensations," *Revue scientifique* 39 (1887): 589.

97. A.-M. Bloch, "Expériences sur la vitesse relative des transmissions visuelles, auditives, et tactiles," *Journal de l'anatomie et de la physiologie normales et pathologiques de l'homme et des animaux* 20 (1884): 26.

98. Jules Tannery, letter to Émile Boutroux, April 1873, MS 4122.2, Cahiers d'étudiant et papiers divers d'Émile Boutroux, Bibliothèque de l'Institut de France.

99. Boutroux, *Contingency of the Laws of Nature*, 123.

100. Ibid., 68.

101. Ibid., 121, 138.

102. Boutroux, *Natural Law in Science and Philosophy*, 183.

103. Ibid., 185.

104. Chevalier, citing Raymond Thamin, in *Entretiens avec Bergson*, 78.

105. Kant, *Critique of Pure Reason*, B50, 77.

106. Henri Bergson, *Time and Free Will*, 232.

107. Bergson explicitly referenced Jules Tannery's critique of psychophysics. See Bergson, *Time and Free Will*, 67. Bergson's debt to Boutroux, however, went unacknowledged in *Time and Free Will*. Laurent Fedi documents the influence of Boutroux on Bergson's first book in "Bergson et Boutroux, la critique du modèle physicaliste et des lois de conservation en psychologie," *Revue de métaphysique et de morale* 2, no. 30 (2001): 97–118.

108. "Entretien avec Bergson," MS 4122.7, Cahiers d'étudiant et papiers divers d'Émile Boutroux, Bibliothèque de l'Institut de France.

109. Dominique Parodi, *Du positivisme à l'idéalisme*, vol. 2 (Paris: J. Vrin, 1930), 139. The sixth chapter, "Émile Boutroux," synthesizes the notes from interviews that Parodi bequeathed to Boutroux's archive at the Institut de France.

110. Émile Boutroux, letter to Henri Lachelier, June 29, 1879, MS 18742, Fonds Henri Lachelier, Bibliothèque nationale de France.

111. Bergson employed several examples of movement in *Time and Free Will*. I am using the runner to reconstruct his argument in the book's second chapter.

112. Victor Egger, "La psychologie physiologique," *Revue philosophique* 5 (1878): 233.

113. Ibid.

114. Charles Richet, "Réplique de M. Richet," *Revue philosophique* 5 (1878): 238, 240.

115. Charles Richet, "Sur la méthode de la psychologie physiologique," *Revue philosophique* 5 (1878): 33.

116. Victor Egger, *La parole intérieure: Essai de psychologie descriptive* (Paris: Félix Alcan, 1881), 113.

117. Victor Egger, Notes et documents de travail, MSVC 419, Papiers Victor Egger, Bibliothèque Victor Cousin, la Sorbonne, Paris.

118. Bergson, *Time and Free Will*, 67–68.

119. Ribot, "De la durée des actes psychiques d'après des travaux récents," 288.

120. Bergson, *Time and Free Will*, 100, translation altered.

121. It has been debated in Bergson scholarship whether lived duration is seamless or separable. Gaston Bachelard (1884–1962) levied a critique of Bergson's concept of duration on the grounds that it forecloses the possibility of analyzing the instants constitutive of time. See Bachelard. *L'intuition de l'instant* (Paris: Félix Alcan, 1932).

122. Bergson, *Time and Free Will*, 73–74.

123. I owe my reading to Gilles Deleuze, "La conception de la différence chez Bergson," *Etudes bergsoniennes* 4 (1956): 77–112.

124. Bergson, *Time and Free Will*, 104–6.

125. Ibid., 128.

126. Alfred Fouillée, "Correspondance," *Revue de métaphysique et de morale* 20, no. 1 (1912): 26.

127. Augustin Guyau, *La philosophie et la sociologie d'Alfred Fouillée* (Paris: Félix Alcan, 1913), 88.

128. Ibid.

129. Ibid.

130. Alfred Fouillée, *La philosophie de Platon: Exposition, histoire et critique de la théorie des idées* (Paris: Librairie philosophique de Ladrange, 1869), 728.

131. James Kloppenberg situates Fouillée in a transatlantic intellectual formation that advanced a "radical theory of knowledge" predicated on a *via media* between metaphysics and science. My interpretation is indebted to Kloppenberg's, although I read conciliation as a method that elevated spiritualism on the basis of experimentation. See Kloppenberg, *Uncertain Victory*.

132. Alfred Fouillée, *La liberté et le déterminisme*, 2nd ed. (Paris: Félix Alcan, 1884), 358.

133. Théodule Ribot, "Philosophy in France," *Mind* 2, no. 7 (1877): 372.

134. Alfred Fouillée, "La mémoire et la reconnaissance des idées," *Revue des deux mondes* 70 (1885): 146.

135. Alfred Fouillée, "La vie consciente et la vie inconsciente d'après la nouvelle psychologie," *Revue des deux mondes* 59 (1883): 904.

136. See Henri Beaunis, *Recherches expérimentales sur les conditions de l'activité cérébrale et sur la physiologie des nerfs* (Paris: J.-B. Baillière, 1884).

137. Marey, "Leçon d'ouverture," 30.

138. Alfred Fouillée, "La philosophie des idées-forces: Comme conciliation du naturalisme et de l'idéalisme," *Revue philosophique* 8 (1879): 7.

139. Alfred Fouillée, letter to Charles Renouvier, May 28, 1879. REN 069.2, Fonds Charles Renouvier, Université Paul Valéry—Montpellier III.

140. Alfred Fouillée, "La vie consciente et la vie inconsciente d'après la nouvelle psychologie," 888.

141. Ibid.

142. Annamaria Contini, "L'intelligence créatrice: Puissance et volonté de conscience dans la philosophie d'Alfred Fouillée," 40.

143. Alfred Fouillée, *L'évolutionnisme des idées-forces* (Paris: Félix Alcan, 1890), liv.

144. Ibid.

145. Ibid., 36.

146. Bergson, *Time and Free Will*, 106–7.

147. Fouillée, *L'évolutionnisme des idées-forces*, lix.

148. Bergson, *Time and Free Will*, 233.

149. Bergson, *Creative Mind*, 22.

150. See Jacqueline Carroy, Annick Ohayon, and Régine Plas, *Histoire de la psychologie en France* (Paris: La Découverte, 2006); Serge Nicolas, *Histoire de la psychologie française: Naissance d'une nouvelle science* (Paris: L'Harmattan, 2002).

151. Théodule Ribot, letter to Paul Tannery, February 26, 1885, in Paul Tannery, *Mémoires scientifiques*, vol. 16 (Paris: Cauthier-Villars, 1939), 260.

152. See Raymond E. Fancher, *The Intelligence Men: Makers of the IQ Controversy* (New York: W. W. Norton, 1987).

153. See M. Huteau, *Psychologie, psychiatrie et société sous la troisième république: La biocratie d'Édouard Toulouse (1856–1947)* (Paris: L'Harmattan, 2002).

154. See Laura L. Koppes, ed., *Historical Perspectives in Industrial and Organizational Psychology* (New York: Psychology Press, 2007).

Chapter Three

1. Course notes written by Eugène Estival and published in Jean Brady, *Bergson Professeur: Au lycée de Clermont-Ferrand, Cours 1885–1886* (Paris: L'Harmattan, 1998), 65.

2. Ibid.

3. Philippe Soulez, Bergson's biographer, warns that it would be "premature to make the published works and the courses into two inseparable halves of the Bergsonian œuvre." See Soulez and Worms, *Bergson*, 62.

4. Cited in Jean Guitton, *La vocation de Bergson* (Paris: Gallimard, 1960), 67.

5. Gutting, *French Philosophy in the Twentieth Century*, 9.

6. Jan Goldstein argues that Cousin left an indelible imprint on the human sciences in France into the twentieth century in *Post-Revolutionary Self*. Similarly, John Brooks III, *Eclectic Legacy*, argues that Cousin's legacy endured despite the anti-spiritualist pretensions of scientific thinkers such as Alfred Binet, Émile Durkheim, and Théodule Ribot.

7. Although the content in the program for the contemporary philosophy course has changed, it retains a similar organization. Its five sections, in order, are the subject (with lessons in consciousness, perception, the unconscious, the other, desire, and existence and time), culture, reason and reality, politics, and ethics. For an exemplary guide, see Frédéric Milon et al., *Objectif bac: Toutes les matières terminale* (Paris: Hachette, 2019).

8. My argument runs contrary to that of Goldstein, who claims that Cousin's influence persisted well beyond his death: "As a regime that embraced scientific positivism and an active anticlerical policy once it became fully 'republicanized' around 1880 . . . the early Third Republic would seem to have had every reason to unseat the old

Cousinian philosophy." This was not the case, she elaborates, because "psychology, still presented as the first and foundational branch of philosophy, still operated with a tripartite consciousness comprised of sensation, reason, and will." Jan Goldstein, "Neutralizing Freud: The Lycée Philosophy Class and the Problem of the Reception of Psychoanalysis in France," *Critical Inquiry* 40, no. 1 (2013): 53. To the contrary, I argue that although Cousin's mark was not expunged from the psychology section, the philosophy course nevertheless underwent a significant rupture with Cousin's eclectic legacy during the final decades of the nineteenth century.

9. See Eugen Weber, *Peasants into Frenchmen: The Modernization of Rural France 1870–1914* (Stanford, CA: Stanford University Press, 1976), 303–38.

10. Philip Nord, *The Republican Moment: Struggles for Democracy in Nineteenth-Century France* (Cambridge, MA: Harvard University Press, 1995), 32.

11. "Concours de 1881," *Bulletin Administratif du ministère de l'instruction publique* 24 (October 31, 1881): 810. The grueling examination consisted of appreciating and critiquing three Greek texts, two in Latin, and two in modern French, in addition to delivering an oral lesson. Students spent a year intensively studying the texts chosen by the jury. Alan Schrift contends that the agrégation played a profound role in shaping the ideas of French philosophers. See Schrift, "The Effects of the *Agrégation de philosophie* on Twentieth-Century French Philosophy," *Journal of the History of Philosophy* 46, no. 3 (2008): 449–74.

12. On the emergence of the meritocratic intellectual as a social category under the Third Republic, see Prochasson, *Les années électriques, 1880–1910*; Christophe Charle, *Naissance des "intellectuels" (1880–1900)* (Paris: Éditions de minuit, 1990).

13. *Annuaire de l'instruction publique et des beaux-arts pour l'année 1882, première partie: Administration et personnel* (Paris: Delalain, 1882), 475.

14. Chevalier, *Entretiens avec Bergson*, 19. The same transformation is recounted by Joseph Desaymard, a student of Bergson, in "H. Bergson à Clermont-Ferrand," in *Bulletin historique et scientifique d'Auvergne* (Clermont-Ferrand: Bellet, 1910): 216.

15. Christophe Charle, *Les professeurs de la faculté des lettres de Paris, dictionnaire biographique, volume 1 (1809–1908)* (Paris: CNRS, 2000).

16. Christophe Charle, *La république des universitaires, 1870–1940* (Paris: Éditions du Seuil, 1994), 191.

17. Bergson was also a replacement instructor for the chair of Greek and Latin at the Collège de France from 1896 to 1900.

18. My account of Bergson's ascendency diverges from much sociological scholarship. Jean-Louis Fabiani extensively documents the institutions—including the *agrégation*, *programme*, and *baccalauréat*—that mediated the intellectual field in which "the philosophers of the Republic" wielded their cultural capital. See Fabiani, *Les philosophes de la république* (Paris: Éditions de minuit, 1988). On the notion of an intellectual field, see Pierre Bourdieu, "Intellectual Field and Creative Project," *Social Science Information* 8 (1969): 89–119. Although sociological methods help historians grasp the changing role of philosophers as a class, these methods often neglect individual thinkers' lives and the shifting meanings of their ideas. Moreover, most sociological approaches situate French philosophers in a narrow geography. Perched atop the hill in the Latin Quarter, the University of Paris system figures as the central node in these studies of the academic circuit wherein professors and students jockeyed for position. This chapter widens the geographic focus. I examine the distinct experience that teaching in the lycées conferred on young professors still distanced from the

Parisian universities. For sociological histories of French intellectual culture in the fin de siècle, see Fritz Ringer, *Fields of Knowledge: French Academic Culture in Comparative Perspective, 1890–1920* (New York: Cambridge University Press, 1992); Louis Pinto, *Les philosophes entre le lycée et l'avant-garde, les métamorphoses de la philosophie dans la France aujourd'hui* (Paris: L'Harmattan, 1987); Christophe Charle, "Le champ universitaire parisien à la fin du XIXe siècle," *Actes de la recherché en sciences sociales* 8 (1983): 47–48.

19. Henri Marion, "Le nouveau programme de philosophie," in *Les philosophes saisie par l'état*, ed. Stéphane Douailler et al. (Paris: Aubier-Montaigne, 1992), 516; originally published in *Revue philosophique* 10 (1880): 414–28.

20. Paul Janet in *Pour et contre l'enseignement philosophique*, ed. F. Vandérem (Paris: Félix Alcan, 1894), 54.

21. Alfred Binet, "Enquête sur l'évolution de l'enseignement," *L'année psychologique* 14 (1907): 167.

22. Ibid., 169.

23. Ibid.

24. Weisz, *Emergence of Modern Universities in France*, 9.

25. See Isabelle Havelange, Françoise Huguet, and Bernadette Lebedeff, *Les inspecteurs généraux de l'instruction publique: Dictionnaire biographique, 1802–1914* (Paris: Institut National de Recherche Pédagogique, 1986).

26. See Jules Simon, *Victor Cousin* (Paris: Hachette, 1887).

27. Jules Simon, *Réforme de l'enseignement secondaire* (Paris: Hachette, 1874), 13; Michel de Montaigne, *The Complete Essays of Montaigne*, trans. Donald M. Frame (Stanford, CA: Stanford University Press, 1958), 100.

28. Simon, *Réforme de l'enseignement secondaire*, 77.

29. Bruno Poucet, *Enseigner la philosophie: Histoire d'une discipline scolaire, 1860–1990* (Paris: CNRS, 1999), 136.

30. On Ferry's endorsement of Comte, see Claude Nicolet, *L'idée républicain en France (1789–1924)* (Paris: Gallimard, 1982), 256–57.

31. See Alfred Fouillée, "La réforme de l'enseignement philosophique et morale en France," *Revue des deux mondes* (1880): 332–69.

32. Francisque Bouillier, *L'université sous M. Ferry* (Paris: Gaume, 1880), 289.

33. Marion, "Le nouveau programme de philosophie," 510.

34. Jean-Louis Fabiani suggests that professors' newfound autonomy made it possible for works of celebrity philosophers to emerge, from Bergson's *Matter and Memory* through Jean-Paul Sartre's *Being and Nothingness*. Jean-Louis Fabiani, *Qu'est-ce qu'un philosophe français?* (Paris: Éditions de l'École des hautes études en sciences sociales, 2010), 53.

35. Paul Janet, "Rapport sur l'enseignement de la philosophie secondaire" (1880) reproduced in Jean-Louis Dumas, "L'enseignement de la philosophie en 1880," *Revue de l'enseignement philosophique* 25 (1975): 57.

36. Henri Marion elaborated these principles in *Education dans l'université* (Paris: Armand Colin, 1891). Marion left the codification of pedagogy up to the philosopher and director of primary education Ferdinand Buisson, who published the *Dictionnaire de pédagogie et d'instruction primaire* (Paris: Hachette, 1887). In 1890, Buisson took over Marion's chair at the Sorbonne.

37. In defense of the law, Sée founded the journal *L'enseignement secondaire des jeunes filles* in 1881.

38. Françoise Mayeur, *L'enseignement secondaire des jeunes filles sous la troisième république* (Paris: Fondation nationale des sciences politiques, 1977), 13.

39. Jules Ferry, *Discours et opinions de Jules Ferry*, vol. 1, ed. Paul Robiquet (Paris: Armand Colin, 1893), 304.

40. See *Revue de l'enseignement secondaire et de l'enseignement supérieur* (1885): 508–9.

41. *Bulletin Administratif du ministère de l'instruction publique* 37 (1885): 213–16.

42. Émile Boutroux in *Pour et contre l'enseignement philosophique*, ed. F. Vandérem et al. (Paris: Félix Alcan, 1894), 33.

43. Ibid.

44. Théodule Ribot in ibid., 28.

45. Alfred Fouillée in ibid., 70.

46. See Alexandre Ribot, *Enquête sur l'enseignement secondaire* (Paris: Belin Frères, 1899). The report also included the notable spiritualist philosophers Gustave Belot and Félix Ravaisson.

47. These were divided into tracks A (Latin and Greek); B (Latin and modern languages); C (Latin and sciences); and D (modern languages and sciences).

48. Alfred Fouillée, *La conception morale et civique de l'enseignement* (Paris: Éditions de la Revue Bleue, 1902), 27.

49. Gebhart, *La voix nationale*, April 7, 1902.

50. *Bulletin Administratif du ministère de l'instruction publique* 72 (1902): 760–62.

51. Louis Liard, *Le nouveau plan d'études de l'enseignement secondaire* (Paris: Cornély, 1903), 11.

52. See Anne-Marie Drouin-Hans, "Sciences naturelles et philosophie: Les enjeux d'un territoire conceptual," in *Sciences naturelles et formation de l'esprit: Autour de la réforme de l'enseignement de 1902*, ed. Nicole Hulin (Lille: Septentrion, 2002), 107–28.

53. Alfred Fouillée, *L'enseignement au point de vue national*, 2nd ed. (Paris: Hachette, 1909), 145.

54. Poucet, *Enseigner la philosophie*, 70.

55. Paul Gerbod, *La vie quotidienne dans les lycées et collèges au XIXe siècle* (Paris: Hachette, 1968), 16.

56. Émile Boirac, *La dissertation philosophique* (Paris: Félix Alcan, 1890), xvii.

57. François Evellin, "La philosophie au lycée—Vocabulaire, méthode, enseignement," *Revue de l'enseignement secondaire et de l'enseignement supérieur* 4 (1884): 164.

58. Ibid., 170.

59. The headmaster of Lycée Blaise Pascal recounted: "Mr. Evellin, inspector general of philosophy classes, inspected Mr. Bergson's class. He expressed to me his complete satisfaction that is was well managed and praised the professor, whose excellent lesson on the *Novum organum* he listened to with interest." Cited in Henri Bergson, *Leçons Clermontoises I*, ed. Renzo Ragghiani (Paris: L'Harmattan, 2003), 10–11.

60. Cited in Chevalier, *Entretiens avec Bergson* (Paris: Plon, 1959), 178.

61. From Bergson's dossier at the Archives Nationales. Cited in Soulez and Worms, *Bergson*, 88.

62. "Cours de Philosophie: Notes prises par René Waltz à Lycée Henri IV 1893," Cahier IV, 307, Papiers Henri Bergson, MSVC 307, Bibliothèque Victor Cousin, la Sorbonne, Paris.

63. Ibid.

64. Jules Lagneau, "Traité de dieu, de l'homme et de la béatitude de Spinoza,

P. Janet," *Revue philosophique* 7 (1879): 67–79; Lagneau, "De la métaphysique, sa nature et ses droits dans ses rapports avec la religion et avec la science, pour servir d'introduction à la métaphysique d'Aristote, Barthélemy Saint-Hilaire," *Revue philosophique* 9 (1880): 210–35; Lagneau, "*Spinoza, His Life and Philosophy*, Fr. Pollock," *Revue philosophique* 13 (1882): 306–15.

65. André Canivez, "Jules Lagneau, professeur et philosophe: Essais sur la condition du professeur de philosophie jusqu'à la fin du XIXe siècle" (PhD diss., Université de Strasbourg, 1965), 368.

66. Comments made by Amédée-Thierry, a student of Langeau in 1892–93, in ibid., 371–72.

67. Notes taken by Mr. Lejoindre. See Jules Lagneau, *Cours de psychologie 1886–87*, vol. 2, *La psychologie: Objet, méthode, et division*, ed. Emmanuel Blondel (Dijon: RDP de Bourgogne, 1997), 24.

68. Ibid., 47.

69. Ibid., 54.

70. "Rapports des Inspecteurs généraux de 1879," AN F17 22934, Archives nationales, Paris.

71. Cited in Jules Lagneau, *Cours intégral 1886–87*, vol. 1, *Histoire de la philosophie*, ed. Emmanuel Blondel (Dijon: RDP de Bourgogne, 1996), 15.

72. Cited in Canivez, "Jules Lagneau," 364.

73. Ibid.

74. Notes taken by Joseph Desaymard in Henri Bergson, *Cours I. Leçons de psychologie et de métaphysique: Clermont-Ferrand, 1887–1888*, ed. Henri Hude (Paris: Presses universitaires de France, 1990), 33.

75. Wilhelm Waldeyer, "Über einige neuere Forschungen im Gebiete der Anatomie des Zentralnervensystems," *Deutsche Medizinische Wochenschrift* 17, no. 50 (1891): 1352–56.

76. See Gordon M. Shepherd, *Foundations of the Neuron Doctrine* (New York: Oxford University Press, 1991).

77. In England, Charles Scott Sherrington showed that synapses facilitate reflex responses, which activate the integration of information from the peripheral nervous system to the brain and back. Sensory stimulations thereby pass into electrical impulses. See Sherrington, *The Integrative Action of the Nervous System* (New Haven, CT: Yale University Press, 1906).

78. The German neurologist Joseph von Gerlach (1820–96) conceptualized nervous fibers as interconnected protoplasm. See Joseph von Gerlach, "Von dem Rückenmarke," in *Lehre von den Geweben des Menschen und der Thiere*, ed. Salomon S. Stricker (Leipzig: Engelmann, 1871), 663–93. This material was thought to be made of dense tissues of thin filaments. For the Italian neurologist Camillo Golgi (1843–1926), the nervous system constituted a "diffuse nerve network." See Camillo Golgi, "The Neuron Doctrine—Theory and Facts; Nobel Lecture, December 11, 1906," in *Nobel Lectures: Physiology or Medicine 1901–1921* (Amsterdam: Elsevier, 1967), 189–217.

79. Santiago Ramón y Cajal, "Sobre las fibras nerviosas de la capa molecular del cerebelo," *Revista trimestral de histología normal y patológica* 1 (1888): 33–49.

80. Brady, *Bergson Professeur*, 69.

81. Bergson recounts having delved into psychophysics because "it was on the order of the day and, in the field of an examination Fechner's theory, I had the chance to be understood and followed." Charles Du Bos, *Journal I (1922–23)* (Paris: Corréa,

1946), 63–68; reproduced in Henri Bergson, *Écrits et paroles*, vol. 2, ed. R. M. Mossé-Bastede (Paris: Presses universitaires de France, 1957), 238–40.

82. "Notes de cours prises par E. Chartier (1888)," NAF 17718, Papiers d'Alain, Bibliothèque Nationale de France.

83. See Marcel Conche, "Bergson à Clermont," *L'enseignement philosophique* 47, no. 2 (1996): 6.

84. Each reissue offered little new content. Reissues did give the publishing house, Hachette, the opportunity to insert new advertisements for other manuals.

85. Élie Rabier, *Du rôle de la philosophie dans l'éducation* (Paris: Delalain, 1886), 4.

86. See Binet, "Enquête sur l'évolution de l'enseignement," 213.

87. Clippings from Élie Rabier, *Leçons de philosophie: I Psychologie* (Paris: Hachette, 1884), 49–57, NAF 16611, Fonds Marcel Proust, Bibliothèque nationale de France.

88. Rabier's letters at the Institut de France and Bibliothèque nationale de France are peppered with notices sent to philosophy professors who received a copy of his manual. The publishing house, Hachette, maintained a list of professors who were sent free copies.

89. Alain, *Souvenirs concernant Jules Lagneau* (Paris: Gallimard, 1925), 722.

90. Rabier, *Leçons de philosophie*, 68.

91. Bergson, *Cours I. Leçons de psychologie et de métaphysique*, 91.

92. Henri Bergson, *Cours de psychologie de 1892–1893 au lycée Henri-IV*, ed. Sylvain Matton (Paris: SÉHA, 2008), 162.

93. Ibid.

94. Henri Bergson, "La place et le caractère de la philosophie dans l'enseignement secondaire Séance du 18 décembre 1902," in *Mélanges*, ed. A. Robinet (Paris: Presses universitaires de France, 1972), 568–69.

95. The Ministry of Public Instruction reserved the right to survey and censor manuals following the reinstatement of philosophy instruction in 1864 under the Second Empire. But the ministry never told professors which manuals to adopt. The responsibility of surveying manuals was left to local administrations, which kept few records of the manuals professors used. It was ultimately up to professors to choose their own manuals.

96. See Alain Choppin, *Les manuels scolaires: Histoire et actualité* (Paris: Hachette, 1992). The publication of manuals in all lycée subjects escalated during the period. While in 1867 only 188 manuals were published, by 1873 there were 602; and in 1883 the number of manuals available leapt to 933.

97. A complete list of philosophy manuals is found in Poucet, *Enseigner la philosophie*, 404–8.

98. There were four types of philosophy textbooks: pedagogical books, which professors used to guide their own teaching; manuals published by and for religious schools; dissertation manuals, which instructed students how to write philosophy essays; and educational manuals used in the terminal class of the public lycées. I focus on the final type.

99. Amedée Jacques, Jules Simon, and Émile Saisset, *Manuel de philosophie à l'usage des collèges* (Paris: Joubert, 1846), vi.

100. Conche, "Bergson à Clermont," 6.

101. Paul Janet, *Traité élémentaire de philosophie à l'usage des classes* (Paris: Delgrave, 1879), vi.

102. Ibid.

103. Paul Janet, *La crise philosophique* (Paris: Germer Baillière, 1865).

104. Janet, *Traité élémentaire de philosophie*, vi.

105. Serge Nicolas, *Études d'histoire de la psychologie* (Paris: L'Harmattan, 2009), 117.

106. Paul Janet, *Le cerveau et la pensée* (Paris: Germer Baillière, 1867), 177.

107. See Brooks, *Eclectic Legacy*, 130.

108. Abel Rey, *Leçons élémentaires de psychologie et de philosophie* (Paris: Édouard Cornély, 1903), i–ii (italics in original).

109. Charles Jourdain, *Questions de philosophie pour l'examen du baccalauréat ès lettres*, 1st ed. (Paris: Hachette, 1847).

110. Charles Jourdain, *Notions de logique*, 6th ed. (Paris: Hachette, 1859).

111. Charles Jourdain, *Notions de philosophie*, 17th ed. (Paris: Hachette, 1882), ii.

112. Charles Jourdain, *L'école sans dieu* (Paris: Jules Gervais, 1880).

113. Ibid., 465.

114. Charles Jourdain, *Notions de philosophie*, 18th ed. (Paris: Hachette, 1888), 476.

115. Fonsegrive went on to publish a complementary manual, which addressed the rest of the program: *Éléments de philosophie: II. Logique, métaphysique, morale, histoire de la philosophie, dissertations philosophiques* (Paris: Alcide Picard et Kaan, 1892).

116. Georges-L. Fonsegrive, *Éléments de philosophie: I. Psychologie* (Paris: Alcide Picard et Kaan, 1891), 28.

117. Ibid., 22.

118. Ibid., 5.

119. Georges-L. Fonsegrive, *Essai sur le libre arbitre* (Paris: Félix Alcan, 1887), 516.

120. Fonsegrive, *Éléments de philosophie*, 5.

121. Bernard Perez, "*Eléments de philosophie: 1. Psychologie* par G.-L. Fonsegrive," *Revue philosophique* 33 (1892): 83.

122. See Georges-L. Fonsegrive, *L'attitude du catholique devant la science*, 2nd ed. (Paris: Librairie Bloud et Barral, 1900); Fonsegrive, *Le Catholicisme et la vie de l'esprit* (Paris: Victor Lecoffre, 1899).

123. Rey, *Leçons élémentaires de psychologie et de philosophie*, 15.

Chapter Four

1. Ribot published numerous works in psychopathology, including treatises on disorders of personality and volition. See Ribot, *Les maladies de la personnalité* (Paris: Félix Alcan, 1885); Ribot, *Les maladies de la volonté* (Paris: Félix Alcan, 1882).

2. Théodule Ribot, *Diseases of Memory: An Essay in the Positive Psychology*, trans. William Huntington Smith (New York: D. Appleton, 1882), 160; originally published as *Les maladies de la mémoire* (Paris: Félix Alcan, 1881).

3. Théodule Ribot, *Les maladies de la mémoire*, 10th ed. (Paris: Félix Alcan, 1895), 11.

4. Henri Bergson, *Matter and Memory*, trans. Margaret Paul and W. Scott Palmer, 8th ed. (Brooklyn, NY: Zone Books, 2008), 30; originally published as *Matière et mémoire* (Paris: Félix Alcan, 1896).

5. The lobes' anatomical organization was established by the mid-nineteenth century. The frontal lobe was shown to control higher order mental functions; the pari-

etal lobes facilitated movement and spatial orientation; the occipital lobe coordinated vision; and the temporal lobe was concerned with auditory stimuli, memory, and speech. Physiological inquiry during the second half of the century revealed the lobes' particular functions. These divisions, although accepted as a basic starting point of neurophysiology, were arbitrary in that numerous interhemispheric functions have since been shown to traverse the boundaries between the lobes. See Edwin Clarke and Kenneth Dewhurst, "The Genesis of Cortical Localization," in *An Illustrated History of Brain Function: Imaging the Brain from Antiquity to the Present* (San Francisco: Norman, 1996), 115–27.

6. Bergson, *Matter and Memory*, 74.

7. See E. Jaffard, B. Claverie, and B. Andrieu, *Cerveau et mémoires: Bergson, Ribot et la neuropsychologie* (Paris: Osiris, 1998); Jean-Noël Massa, "Matérialisme et neurosciences: La question des localisations cérébrales," *Revue philosophique* 185 (1995): 43–53; Anne Harrington, "Au-delà de la phrénologie: Théories de la localisation à l'époque contemporaine," in Pietro Corsi, ed., *La fabrique de la pensée: La découverte du cerveau, de l'art de la mémoire aux neurosciences* (Milan: Electa, 1990), 206–15.

8. Sergueï Korsakoff, "Étude medico-psychologique sur une forme des maladies de la mémoire," *Revue philosophique* 28 (1889): 504.

9. See Timothy Lenoir, *Instituting Science: The Cultural Production of Scientific Disciplines* (Stanford, CA: Stanford University Press, 1997), 75–95.

10. See Scott O Lilienfeld et al., "Fifty Psychological and Psychiatric Terms to Avoid: A List of Inaccurate, Misleading, Misused, Ambiguous, and Logically Confused Words and Phrases," *Frontiers in Psychology* 6, no. 1100 (August 3, 2015), https://doi.org/10.3389/fpsyg.2015.01100.

11. See Mark S. Morrisson, *Modernism, Science, and Technology* (New York: Bloomsbury, 2016); Laura Salisbury and Andrew Shail, eds., *Neurology and Modernity: A Cultural History of Nervous Systems, 1800–1950* (New York: Palgrave, 2010); Andreas Killen, *Berlin Electropolis: Shock, Nerves, and German Modernity* (Berkeley: University of California Press, 2006).

12. Allison Muri, *The Enlightenment Cyborg: A History of Communications and Control in the Human Machine, 1660–1830* (Toronto: University of Toronto Press, 2007); Kyle L. Kirkland, "High-Tech Brains: A History of Technology-Based Analogies and Models of Nerve and Brain Function," *Perspectives in Biology and Medicine* 45, no. 2 (2002): 212–33; Iwan Rhys Morus, "'The Nervous System of Britain': Space, Time and the Electric Telegraph in the Victorian Age," *British Journal for the History of Science* 33 (2000): 455–75; Anson Rabinbach, *The Human Motor: Energy, Fatigue and the Origins of Modernity* (New York: Basic Books, 1990); Georges Canguilhem, "The Role of Analogies and Models in Biological Discovery," in *Scientific Change: Historical Studies in the Intellectual, Social, and Technical Conditions for Scientific Discovery and Technical Invention*, ed. A. C. Crombie (New York: Basic Books, 1963), 507–20.

13. Laura Otis, *Networking: Communicating with Bodies and Machines in the Nineteenth Century* (Ann Arbor: University of Michigan Press, 2001).

14. Friedrich A. Kittler, *Discourse Networks 1800/1900*, 1985, trans. Michael Metteer and Chris Cullens (Stanford, CA: Stanford University Press, 1990).

15. Jean-Marie Guyau, "La mémoire et le phonographe," *Revue philosophique* 9 (1880): 319.

16. Wiring figures also surfaced in literature from the period. See Anne Stiles,

Popular Fiction and Brain Science in the Late Nineteenth Century (Cambridge: Cambridge University Press, 2012).

17. Laurence J. Kirmayer argues that scientific metaphors have "surplus meaning, specific connotations that were not intended by those who created or chose the metaphor to capture some intuitions about the nature of brain functioning and perhaps to build a model on that analogy." Kirmayer, "The Future of Critical Neuroscience," in *Critical Neuroscience: A Handbook of the Social and Cultural Contexts of Neuroscience*, ed. Suparna Choudhury and Jan Slaby (Oxford: Wiley-Blackwell, 2012), 371. To the contrary, I argue that spiritualist thinkers deployed rhetorical figures consciously. They illuminated the vexed relations of spirit and matter in cerebral localization and the psychopathology of memory.

18. Jean-Claude Wartelle, "La Société d'anthropologie de Paris de 1859 à 1920," *Revue d'histoire des sciences humaines* 1, no. 10 (2004): 125–71; Wartelle, *Paul Broca: Founder of French Anthropology, Explorer of the Brain* (New York: Oxford University Press, 1992).

19. See Stephen Jay Gould, *The Mismeasure of Man*, 1981 (New York: W. W. Norton, 1996), 114–41.

20. See Alice L. Conklin, *In the Museum of Man: Race Anthropology, and Empire in France, 1850–1950* (Ithaca, NY: Cornell University Press, 2013).

21. Paul Broca, "Remarks on the Seat of the Faculty of Articulate Language, Followed by an Observation of Aphemia (1861)," in *A Source Book in the History of Psychology*, ed. Edwin G. Boring and Richard Hernstein (Cambridge, MA: Harvard University Press, 1965), 224.

22. See Anne Harrington, *Medicine, Mind, and the Double Brain: A Study in Nineteenth-Century Thought* (Princeton, NJ: Princeton University Press, 1989).

23. Broca, "Remarks on the Seat of the Faculty of Articulate Language," 225.

24. Ibid., my emphasis.

25. Paul Broca, "Localization of Speech in the Third Left Frontal Convolution (1865)," trans. Ennis Berker, Ata Berker, and Aaron Smith, *Archives of Neurology* 43, no. 1 (1986): 1068.

26. Pierre Foissac, *Le matérialisme et le spiritualisme scientifique ou les localisations cérébrales*, 2nd ed. (Paris: Baillière, 1881), 153.

27. Ibid., 123.

28. Gustav Fritsch and Eduard Hitzig, "On the Electrical Excitability of the Cerebrum (1870)," in *Some Papers on the Cerebral Cortex*, trans. Gerhardt von Bonin (Springfield, IL: Thomas, 1960), 95, my emphasis; originally published as "Über die elektrische Erregbarkeit des Grosshirns," *Archiv für Anatomie, Physiologie und wissenschaftliche Medicin* 37 (1870): 300–332.

29. See Laura Otis, "Howled Out of the Country: Wilkie Collins and H. G. Wells Retry David Ferrier," in *Neurology and Literature, 1860–1920*, ed. Anne Stiles (New York: Palgrave, 2007), 27–51.

30. David Ferrier, "Experimental Research in Cerebral Physiology and Pathology," *Journal of Anatomy and Physiology* 8, no. 1 (1873): 152; the full report was published by the same title in *West Riding Lunatic Asylum Medical Reports* 3 (1873): 30–96. Ferrier published his localization maps of the human brain in *The Functions of the Brain* (London: Smith, Elder, 1876).

31. Carl Wernicke, "The Aphasia Symptom-Complex: A Psychological Study on

an Anatomical Basis (1875)," in *Reader in the History of Aphasia: From Franz Gall to Norman Geschwind*, ed. Paul Eling (Amsterdam: John Benjamins, 1994), 69–98; originally published as *Der aphasische Symptomen-komplex: Eine psychologische Studies auf anatomischer Basis* (Breslau: M. Cohn und Weigert, 1874).

32. For a brilliant account of Wernicke's contributions to the interconnected model of brain organization, see Katja Guenther, *Localization and Its Discontents* (Chicago: University of Chicago Press, 2015).

33. Theodor Meynert, *Zur Mechanik des Gehirnbaues* (Vienna: W. Braumüller, 1874); Meynert, "Fragment aus den anatomischen Corollarien und der Physiologie des Vorderhirns," *Jahrbücher für Psychiatrie* 2 (1881): 87.

34. See R. E. Graves, "The Legacy of the Wernicke-Lichtheim Model," *Journal of the History of Neuroscience* 6, no. 1 (1997): 3–20.

35. The debate reached a zenith in 1908. At the French Society for Neurology, Pierre Marie defended a holistic theory of brain functions against the localizationist approach of Jules Joseph Dejerine (1870–1917), who claimed that cerebral lesions in discrete centers brought about different aphasiac disorders. See Maurizio Paciaroni and Julian Bogousslavsky, "Jules Joseph Dejerine versus Pierre Marie," *Frontiers of Neurology and Neuroscience* 29 (2011): 162–69.

36. Carl Wernicke, *An Outline of Psychiatry in Clinical Lectures*, trans. Johns Dennison and Robert Miller (New York: Springer, 2015), 7.

37. See Christoph Hoffman, "Helmholtz Apparatuses: Telegraphy as a Working Model of Nerve Physiology," *Philosophie Sciential* 7 (2003): 129–49.

38. Emil du Bois-Reymond, "Über tierische Bewegung (1851)," in *Reden*, vol. 2, ed. Estelle du Bois Reymond (Leipzig: Veit, 1912), 48.

39. On metaphors of inscription and technology in the neurology of memory, see Douwe Draaisma, *Metaphors of Memory: A History of Ideas about the Mind*, trans. Paul Vincent (New York: Cambridge University Press, 2000).

40. Emil du Bois-Reymond, "On the Time Required for the Transmission of Volition and Sensation through the Nerves," *Croonian Lectures on Matter and Force*, ed. Henry Bence Jones (London: John Churchill and Sons, 1868), 97.

41. Paul Laurencin, *Le télégraphe* (Paris: J. Rothschild, 1877), 392.

42. Charles Richet, *Physiologie des muscles et des nerfs* (Paris: Germer Baillière, 1882), 537.

43. Victor Egger, "La physiologie cérébrale et la psychologie," *Revue des deux mondes* 24 (1877): 197.

44. Ibid.

45. Alfred Fouillée, "Le sentiment de l'effort et la conscience de l'action," *Revue philosophique* 28 (1889): 569.

46. Robert Mayer, "Les conséquences de la théorie de la chaleur," *Revue des cours scientifiques* 7 (1870): 126.

47. Ibid.

48. Frédéric Paulhan, *La physiologie de l'esprit* (Paris: Félix Alcan, 1880), 33, my emphasis.

49. Jules Bernard Luys, *Le cerveau et ses fonctions* (Paris: Germer Baillière, 1876), 7.

50. Henry Charlton Bastian, *The Brain as an Organ of the Mind* (New York: D. Appleton, 1880), 37.

51. Theodor Meynert, *Psychiatry: A Clinical Treatise on Diseases of the Fore-Brain*

Based upon a Study of Its Structure, Functions, and Nutrition, trans. Bernard Sachs, 1884 (New York: G. P. Putnam's Sons, 1885), cited in Katja Guenther, "A Body Made of Nerves—Reflexes, Body Maps, and the Limits of the Self in Modern German Medicine" (PhD diss., Harvard University, 2009), 35.

52. M. Duval, "Nerfs: Anatomie et physiologie du système nerveux," in S. Jaccoud, ed., *Nouveau dictionnaire de médecine et de chirurgie pratique* (Paris: Baillière, 1877), 576.

53. Hippolyte Taine, "Les vibrations cérébrales et la pensée," *Revue philosophique* 3 (1877): 9.

54. J. Rambosson, *Phénomènes nerveux, intellectuels et moraux, leur transmission par contagion* (Paris: Firmin-Didot, 1883), 210.

55. Gabriel Tarde, "Les traits communs de la nature et de l'histoire," *Revue philosophique* 14 (1882): 276.

56. Gabriel Tarde, "L'idée de l'"organisme social,'" *Revue philosophique* 41 (1896): 640.

57. Richet, *Physiologie des muscles et des nerfs*, 565.

58. Julene K. Johnson, Marjorie Lorch, Serge Nicolas, and Amy Graziano, "Jean-Martin Charcot's Role in the 19th Century Study of Aphasia," *Brain* 139, no. 7 (2013): 1662–70.

59. Théodule Ribot, "Les bases intellectuelles de la personnalité," *Revue philosophique* 18 (1884): 442.

60. Ibid.

61. Ibid., 440.

62. Léonce Manouvrier, "La fonction psycho-motrice," *Revue philosophique* 17 (1884): 640.

63. I am indebted to Susan Ashley for this reading of French psychologists' political metaphors.

64. Fouillée, "La mémoire et la reconnaissance des idées," 146.

65. Bergson, *Matter and Memory*, 171.

66. Ribot, *Diseases of Memory*, 50.

67. Ibid., 53.

68. Ibid., 129.

69. Ibid., 65.

70. Ibid., 122.

71. Ibid., 10.

72. Alfred Fouillée, "La survivance et la sélection des idées dans la mémoire," *Revue des deux mondes* 69 (1885): 375.

73. Bergson, *Matter and Memory*, 78.

74. Ibid., 120.

75. Ibid.

76. Ribot, *Diseases of Memory*, 162–63. Although *la parole intérieure* has been rendered as "internal voice" by the translator, "internal monologue" would better fit the French connotation.

77. Ibid., 164.

78. Victor Brochard, "*La parole intérieure, essai de psychologie descriptive* par Victor Egger," *Revue philosophique* 13 (1882): 410.

79. Letter from Jules Lachelier to Victor Egger, December 1, 1881, Bibliothèque Victor Cousin, la Sorbonne, Paris, MSVC 418.11–12. Émile Boutroux also found the

thesis compelling in a letter written to Victor Egger July 3, 1881, Bibliothèque Victor Cousin, la Sorbonne, Paris, MSVC 418.9–10.

80. Wernicke, "The Aphasia Symptom-Complex," 92.

81. Joseph Jules Dejerine, "Sur un cas de cécité verbale avec agraphie suivi d'autopsie," *Comptes rendus de la Société de Biologie* 3 (1891): 197–201.

82. Egger, *La parole intérieure*, 111.

83. Letter from Victor Egger to Charles Renouvier. August 16, 1882, REN 063-1, Fonds Charles Renouvier, Université Paul Valéry—Montpellier III.

84. Egger, *La parole intérieure*, 113.

85. Fouillée, "La survivance et la sélection des idées dans la mémoire," 367.

86. Bergson, "Séance du 2 mai 1901," 53.

87. Henri Bergson, personal copy of Théodule Ribot, *Les maladies de la mémoire*, 148, BGN—II 40, Fonds Henri Bergson, Bibliothèque littéraire Jacques Doucet. Italics in original.

88. Ibid., 68.

89. Fouillée, "La mémoire et la reconnaissance des idées," 134.

90. Jean-Marie Guyau, *La genèse de l'idée du temps* (Paris: Félix Alcan, 1890), 50.

91. Bergson, *Matter and Memory*, 82–83.

92. Henri Bergson, "Memory of the Present and False Recognition," in *Mind-Energy*, trans. H. Wildon Carr (New York: Henry Holt, 1920), 165.

93. Jean-Marie Guyau, *La genèse de l'idée du temps*, 50.

94. Ibid., 70.

95. Paul Ricoeur, "From Kant to Guyau," in *Guyau and the Idea of Time*, ed. John A Michon et al. (Amsterdam: North Holland, 1988), 157.

96. Bergson, *Matter and Memory*, 79.

97. Although Guyau figured the brain as a phonograph, he borrowed the figure from the experimental psychologist Joseph Delboeuf, who used the metaphor to characterize memory: "the soul is a book of phonographic recordings." Delboeuf, "Le sommeil et les rêves: Leurs rapports avec la théorie de la mémoire," *Revue philosophique* 8 (1879): 129–69, 329–56, 413–37, 632–47.

98. Patrick Feaster, "Speech Acoustics and the Keyboard Telephone: Rethinking Edison's Discovery of the Phonograph Principle," *ARSC Journal* 38, no. 1 (Spring 2007): 10–43.

99. Jean-Marie Guyau, "La mémoire et le phonographe," 320.

100. Ibid., 321.

101. Johannes Müller, *Elements of Physiology* (London: Yalor and Walton, 1842); cited in C. U. M. Smith, "Musical Instruments as Metaphors in Brain Science: From René Descartes to John Hughlings Jackson," in F. Clifford Rose, ed., *Neurology of the Arts: Painting, Music, Literature* (London: Imperial College Press, 2004), 191–206.

102. Jean-Marie Guyau, "La mémoire et le phonographe," 322.

103. Alfred Fouillée, *La psychologie des idées-forces*, vol. 1 (Paris: Félix Alcan, 1893), 189.

104. Ibid., 117.

105. Fouillée, "La mémoire et la reconnaissance des idées," 148.

106. Henri Bergson, "Analyse de l'ouvrage de Guyau: *La genèse de l'idée de temps* avec un introduction par Albert [*sic*] Fouillée," *Revue philosophique* 31 (1891): 187. In *Matter and Memory*, Bergson took up Fouillée's insight that there is "preformation of the movements which follow in the movements which precede, a preformation

whereby the part virtually contains the whole, as when each note of a tune learnt by heart seems to lean over the next to watch its execution." *Matter and Memory*, 94, translation altered.

107. See Christopher Beauchamp, *Invented by Law: Alexander Graham Bell and the Patent That Changed America* (Cambridge, MA: Harvard University Press, 2015), 43.

108. "Destruction des fils électriques à New York," *La nature* 29, no. 1 (1891): 261.

109. "Les téléphones à Paris: Société générale des téléphones," *Nature* 2, no. 1 (1882): 163–74.

110. See Vanessa Ogle, *The Global Transformation of Time, 1870–1950* (Cambridge, MA: Harvard University Press, 2015); Peter Galison, *Einstein's Clocks, Poincaré's Maps: Empires of Time* (New York: W. W. Norton, 2004).

111. See Stephen Kern, *The Culture of Time and Space: 1880–1918* (Cambridge, MA: Harvard University Press, 1983), 68–70.

112. Maurice de Fleury, *Medicine and the Mind*, trans. Stacy B. Collins (London: Downey, 1900), 174–75. Originally published as *Introduction à la médecine de l'esprit* (Paris: Félix Alcan, 1898).

113. Ibid., 176.

114. Bergson, *Matter and Memory*, 44–45. For a critique of Bergson's use of the central telephone exchange metaphor, see Jean-Noël Missa, *L'esprit-cerveau: La philosophie de l'esprit à la lumière des neurosciences* (Paris: J. Vrin, 1993), 137–64.

115. Gilles Deleuze, *Cinema 2*, 1985, trans. Hugh Tomlinson and Robert Garcia (New York: Bloomsbury, 2013), 217.

116. Sentence highlighted in Bergson's personal copy of Hippolyte Taine, *De l'intelligence*, vol. 1 (Paris: Hachette, 1870), 324, BGN II 80, Fonds Henri Bergson, Bibliothèque littéraire Jacques Doucet, Paris.

117. The British biologist Arthur Keith elaborated on the similarities between the brain and the telephone switchboard in *The Engines of the Human Body* (Philadelphia: Lippincott, 1920), 252–67.

118. Alfred Maury, *Le sommeil et les rêves: Études psychologiques sur ces phénomènes et les divers états qui s'y rattachent, suivies de recherches sur le développement de l'instinct et de l'intelligence dans leurs rapports avec le phénomène du sommeil* (Paris: Didier et Cie, 1861).

119. In his personal copy of Ribot's *Diseases of Memory*, Bergson marked the author's reference to Maury and underlined the passage "absolutely nothing is lost in memory." Bergson's marginalia include a note to self that he should read Maury's book. Henri Bergson, personal copy of *Les maladies de la mémoire*, 148, BGN—II 40, Fonds Henri Bergson, Bibliothèque littéraire Jacques Doucet.

120. Bergson, *Matter and Memory*, 165.

121. For his subtractive theory of memory, Bergson acknowledged his debt to the spiritualist philosopher Félix Ravaisson: "We cannot see how memory could settle within matter; but we do clearly understand how—according to the profound saying of a contemporary philosopher—materiality begets oblivion." Ibid., 177.

122. Ibid., 166.

123. Ibid., 170.

124. Ibid., 95.

125. Ibid., 105.

126. Ibid., 103.

127. Ibid., 165–66.

128. Ibid.

129. Bergson, "Memory of the Present and False Recognition," 157.

130. Bergson, *Matter and Memory*, 127.

131. T. O'Conor Sloane, *The Standard Electrical Dictionary* (New York: Norman W. Hensley, 1892), 234.

132. Bergson, *Matter and Memory*, 105, my emphasis.

133. T. O'Conor Sloane, *Standard Electrical Dictionary*, 234.

134. Bergson, *Matter and Memory*, 9.

135. Ibid., 222.

136. Ibid.

137. Bergson, "Le parallélisme psychophysique et la métaphysique positive," *Bulletin de la Société française de philosophie* 1, no. 2 (1901): 32.

138. Henri Bergson, *Creative Evolution*, trans. Arthur Mitchell (Mineola, NY: Dover, 1998), 277; originally published as *L'évolution créatrice* (Paris: Félix Alcan, 1907).

139. Bergson, *Creative Mind*, 57.

140. Henri Bergson, "L'âme humaine, 1916," in *Mélanges*, ed. A. Robinet (Paris: Presses universitaires de France, 1972), 1210, 1200–1215.

Chapter Five

1. Jacob, "La philosophie d'hier et celle d'aujourd'hui," 170.

2. Ibid., 171.

3. Ibid., 177.

4. Ibid.

5. See Sebastian Luft and Fabien Capeillères, "Neo-Kantianism in Germany and France," in *The History of Continental Philosophy*, ed. Alan D. Schrift (Durham, UK: Taylor and Francis, 2010), 47–85.

6. See Michel Espagne, *En deçà du Rhin* (Paris: Cerf, 2005); Warren Schmaus, "Kant's Reception in France: Theories of the Categories in Academic Philosophy, Psychology, and Social Science," *Perspectives in Science* 11, no. 1 (2003): 3–34; Dominique Bourel and François Azouvi, *De Königsberg à Paris: La réception de Kant en France (1788–1804)* (Paris: Vrin, 1991).

7. Henri Bergson, "Mémoire et reconnaissance," *Revue philosophique de la France et de l'étranger* 41 (1896): 225–48, 380–99.

8. Baptiste Jacob, *Lettres d'un philosophe* (Paris: Édouard Cornély, 1911), 132.

9. Xavier Léon had sought an author to pen a critique of *Matter and Memory*; he offered the opportunity to Jacob after the philosopher Victor Delbos (1862–1916) wrote a conciliatory review of the book, "*Matière et mémoire* par Henri Bergson," *Revue de métaphysique et de morale* 5, no. 3 (1897): 353–89. For more on the decision process leading to Jacob's publication, see Stéphan Soulié, *Les philosophes en république: L'aventure intellectuelle de la Revue de métaphysique et de morale et de la Société française de philosophie (1891–1914)* (Rennes: Presses universitaires de Rennes, 2009), 264–66.

10. Jacob, "La philosophie d'hier et celle d'aujourd'hui," 201.

11. In their biography of Bergson, Philippe Soulez and Frédéric Worms agree that the thrust of Jacob's critique was directed at Bergson's claim to have "collapsed, with

his critique of psychological atomism, any legislation of the understanding." Soulez and Worms, *Bergson*, 86.

12. See Jean-Fabien Spitz, *Le moment républicain en France* (Paris: Éditions Gallimard, 2005).

13. On the politics of the journal's founders, see Eric Brandom, "Liberalism and Rationalism at the *Revue de métaphysique et de morale*, 1902–1903," *French Historical Studies* 39, no. 4 (2016): 749–80.

14. "Introduction," *Revue de métaphysique et de morale* 1 (1893): 1–5.

15. Henri Bergson, "Perception et matière," *Revue de métaphysique et de morale* 4, no. 3 (1896): 257–79.

16. Bergson, "Séance du 2 mai 1901."

17. Le Roy, "Un positivisme nouveau," 139.

18. Alfred Fouillée, "La doctrine de la vie chez Guyau," *Revue de métaphysique et de morale* 14, no. 4 (1906): 515.

19. "Nécrologie: A. Darlu," *Revue de métaphysique et de morale*, April–July, supplement (1921): 2.

20. *Journal officiel de la république française* 20, no. 206 (1888): 3301.

21. Élie Halévy to Xavier Léon (August 31, 1891), in Élie Halévy, *Correspondance 1891–1937* (Paris: Éditions de Fallois, 1996), 65.

22. Xavier Léon to Octave Hamelin (April 4, 1893); cited in Soulié, *Les philosophes en république*, 214.

23. Stéphan Soulié, "Xavier Léon, philosophe," *Archives juives* 39, no. 1 (2006): 143–47; Xavier Léon, *La philosophie de Fichte: Ses rapports avec la conscience contemporaine* (Paris: Félix Alcan, 1902); Léon, *Fichte et son temps*, vols. 1–3 (Paris: Armand Colin, 1922–24).

24. I owe my understanding of Halévy's work to Steven Vincent. See "Élie Halévy and French Socialist Liberalism," *History of European Ideas* 44, no. 1 (2018): 75–97.

25. Léon Brunschvicg, "Spiritualisme et sens commun," *Revue de métaphysique et de morale* 5, no. 5 (1897): 539.

26. Joel Revill, "Taking France to the School of the Sciences: Léon Brunschvicg, Gaston Bachelard, and the French Epistemological Tradition" (PhD diss., Duke University, 2006).

27. My italics are meant to preserve the original French; cited in Dominique Marllié, "Les rapports entre la *Revue de métaphysique et de morale* et la *Revue philosophique*: Xavier Léon et Théodule Ribot; Xavier Léon et Lucien Lévy-Bruhl," *Revue de métaphysique et de morale* 98, nos. 1–2 (1993): 80.

28. Ibid.

29. Charles Renouvier and François Pillon, *Critique philosophique* 2 (1872): 13.

30. M. Laurent Fedi, "Philosopher et républicaniser: La critique philosophique de Renouvier et Pillon, 1872–1889," *Romantisme* 115 (2002): 65–82.

31. Immanuel Kant, *Critique of Pure Reason*, trans. Norman Kemp Smith (Boston: Palgrave, 2003), 136.

32. Numerous philosophers have noted Kant's use of legal metaphors. See Ian Proops, "Kant's Legal Metaphor and the Nature of a Deduction," *Journal of the History of Philosophy* 41, no. 2 (2003): 209–29; Onora O'Neill, *Constructions of Reason* (Cambridge: Cambridge University Press, 1989).

33. Charles Renouvier, *Essais de critique générale, Deuxième essai: L'homme; La raison, la passion, la liberté, la certitude, la probabilité morale* (Paris: Ladrange, 1859),

547; cited in William Logue, *Charles Renouvier: Philosopher of Liberty* (Baton Rouge: Louisiana State University Press, 1993), 31.

34. Renouvier's thought had a significant impact on William James. See Donald Wayne Viney, "William James on Free Will: The French Connection," *History of Philosophy Quarterly* 14, no. 1 (1997): 29–52.

35. Beurier, "M. Renouvier et le criticisme français," *Revue philosophique* 3 (1877): 332. The author described Renouvier's thought as "French criticism." Although Renouvier enjoyed a niche audience, it so happened that he received due recognition as the pioneer of French neo-Kantianism somewhat belatedly thanks to the article.

36. Octave Hamelin, *La système de Renouvier* (Paris: J. Vrin, 1927), 228; cited in Logue, *Charles Renouvier*, 94.

37. Lionel Dauriac to Charles Renouvier, Paris, December 30, 1896, Fonds Charles Renouvier, REN 054-4, Université Paul Valéry—Montpellier III.

38. Ibid.

39. Octave Hamelin à Renouvier, December 31, 1889, Fonds Charles Renouvier, REN 081-2, Université Paul Valéry—Montpellier III.

40. Quoted in Soulié, *Les philosophes en république*, 61.

41. Stéphan Soulié, "*La Revue de métaphysique et de morale* et les congrès internationaux de philosophie (1900–1914): Une contribution à la construction d'une internationale philosophique," *Revue de métaphysique et de morale* 84, no. 4 (2014): 467–81.

42. Émile Boutroux, "Introduction: Allocution de M. Boutroux," *Bibliothèque du Congrès international de philosophie* 1 (1900): xii.

43. Ibid., xiv–xv.

44. Charles Baskerville, "International Congresses," *Science* 32, no. 828 (1910): 652–59.

45. See Congress proceedings in *Revue de métaphysique et de morale* 8, no. 5 (1900): 503–698.

46. Soulié, "*La Revue de métaphysique et de morale* et les congrès internationaux de philosophie (1900–1914)," 471.

47. Louis Couturat, "Lettres à Brunetière, sur le pacifisme de Kant," *Le temps*, March 27 and April 1, 1899.

48. See Louis Couturat, *Pour la langue internationale* (Coulommiers: Paul Brodard, 1906).

49. Fabian de Kloe, "Beyond Babel: Esperanto, Ido and Louis Couturat's Pursuit of an International Scientific Language," in *Information beyond Borders: International Cultural and Intellectual Exchange in the Belle Époque*, ed. W. Boyd Rayward (London: Routledge, 2016), 109–22, esp. 112.

50. *Journal des débats* (October 26, 1928); quoted in Soulié, *Les philosophes en république*, 467.

51. On the tensions between nationalism and internationalism among philosophers and scientists in the early twentieth century, see Larry S. McGrath, "Intellectual Ambassadors: Building Peace across the Atlantic in the Early Twentieth Century," *History of Humanities* 3, no. 1 (2018): 159–75.

52. Xavier Léon, "Séance du 7 février 1901—à la Sorbonne," *Bulletin de la Société française de philosophie* 1 (1901): 3.

53. Lalande's initiative was not the first French philosophical dictionary. Edmond Goblot published *Le vocabulaire philosophique* (Paris: Armand Colin, 1901), but it

did not reflect the collective deliberation that took place in the French Society of Philosophy.

54. André Lalande, "Séance du 23 mai 1901—Propositions concernant l'emploi de certains termes philosophiques," *Bulletin de la Société française de philosophie* 1 (1901): 87.

55. *Âme* means "soul," and *Zététique* refers to the method of inquiry associated with ancient Pyrrhonian skepticism.

56. Lalande, "Séance du 23 mai 1901," 98.

57. Ibid.

58. Ibid., 86.

59. André Lalande, *Vocabulaire technique et critique de la philosophie*, vols. 1–2 (Paris: Presses universitaires de France, 1926).

60. Léon Brunschvicg, "L'idée critique et la système Kantien," *Revue de métaphysique et de morale* 31, no. 2 (1924): 153.

61. Léon Brunschvicg and Élie Halévy, "L'année philosophique 1893," *Revue de métaphysique et de morale* 2, no. 4 (1894): 473–96, 563–90, quote p. 496.

62. Ibid., 484.

63. Ibid., 486.

64. See Frédéric Worms, "Entre critique et métaphysique: La science chez Bergson et Brunschvicg," in *Les philosophes et la science*, ed. Pierre Wagner (Paris: Gallimard, 2002), 403–46.

65. "La notion de l'esprit dans la philosophie contemporaine: Cours professé au lycée Henri IV, 1903–04," MS 6958, 239, Fonds Louis Hautecoeur, Bibliothèque de l'Institut de France.

66. Ibid.

67. Ibid., 239–41.

68. In the transcendental deduction of the *Critique of Pure Reason*, Kant posited the table of categories, consisting of the fundamental concepts underlying our claims to knowledge. These were: Of Quantity (unity, totality, plurality); Of Quality (reality, negation, limitation); Of Modality (possibility-impossibility, existence-nonexistence, necessity-contingency); Of Relation (inherence and subsistence, causality and dependence, community).

69. Brunschvicg and Halévy, "L'année philosophique 1893," 491.

70. Fouillée, *La conception morale et civique de l'enseignement*, 151.

71. Bergson, "Psychophysical Parallelism and Positive Metaphysics," 59.

72. Ibid., 68.

73. Ibid., 59.

74. The idea of a "genetic method" was central to Gilles Deleuze's reading in *Le bergsonisme* (Paris: Presses universitaires de France, 1966).

75. Édouard Le Roy, "Sur quelques objections adressées à la nouvelle philosophie," *Revue de métaphysique et de morale* 9, no. 3 (1901): 316–17.

76. Henri Poincaré's early essays were a formidable influence on the young Le Roy: Poincaré, "Les géométries non euclidiennes," *Revue générale des sciences pures et appliquées* 2 (1891): 769–74; Poincaré, "L'espace et la géométrie," *Revue de métaphysique et de morale* 3, no. 6 (1895): 631–46.

77. Étienne Gilson, "Souvenir de Bergson," *Revue de métaphysique et de morale* 64, no. 2 (1959): 136.

78. Édouard Le Roy, "Science et philosophie (I)," *Revue de métaphysique et de morale* 7, no. 4 (1899): 384.

79. Alfred Fouillée, "Correspondance," *Revue de métaphysique et de morale* 20, no. 1 (1912): 26–28.

80. Alfred Fouillée, *La psychologie des idées-forces*, vol. 1 (Paris: Félix Alcan, 1893), 338.

81. Ibid., 338–39.

82. Léon Brunschvicg, *La modalité du jugement* (Paris: Félix Alcan, 1897); *Introduction à la vie de l'esprit* (Paris: Félix Alcan, 1900).

83. Léon Brunschvicg, "La philosophie nouvelle et l'intellectualisme," *Revue de métaphysique et de morale* 9, no. 4 (1901): 437–38.

84. Ibid.

85. Édouard Le Roy, "Science et philosophie (II)," *Revue de métaphysique et de morale* 7, no. 5 (1899): 544.

86. Le Roy, "Sur quelques objections adressées à la nouvelle philosophie," 311.

87. Henri Bergson, "Introduction to Metaphysics," in *Creative Mind*, 136; originally published as "Introduction à la métaphysique," *Revue de métaphysique et de morale* 11, no. 1 (1903): 1–36.

88. Ibid., 135.

89. Ibid.

90. Bergson, *Time and Free Will*, 233.

91. Bergson, "Introduction to Metaphysics," 155.

92. Ibid., 147.

93. As tempting as I find it to call scientific systems "paradigms," the term sullies the pre-Kuhnian worldview of Bergson's time.

94. Bergson, "Introduction to Metaphysics," 155.

95. Ibid., 161.

96. Léon Brunschvicg, *Modalité du jugement* (Paris: Félix Alcan, 1897), 2.

97. M. Cantecor, "La philosophie nouvelle et la vie de l'esprit," *Revue philosophique* 28 (1903): 255.

98. Alfred Fouillée, *La pensée et les nouvelles écoles anti-intellectualistes*, 2nd ed. (Paris: Félix Alcan, 1911), 347.

99. Kant, *Critique of Pure Reason*, 90.

100. Bergson, "Introduction to Metaphysics," 166–67.

101. Henri Bergson, "Philosophical Intuition," in *Creative Mind*, 91.

102. Bergson, *Creative Evolution*, 10.

103. Ibid., 239.

104. Steven Jay Gould makes this point in *Ever since Darwin* (New York: W. W. Norton, 1977).

105. The eye was not the only instance of convergent evolution that interested Bergson. He also examined sexual reproduction in plants and animals (*Creative Evolution*, 51). Reproductive cells (gametes) form by cutting their respective chromosomes in half (a process called meiosis). When the nuclei fuse during fertilization, whether in plants or mammals, they impart a shared genetic profile to the zygote. Similarly, Bergson explored heteroblasty, a phenomenon first noted by zoologists in the late nineteenth century, whereby plants grow abruptly, as when the narrow leaves of young trees rapidly transform into the thick, rounded leaves of maturing

trees (ibid., 75). These heteroblastic transformations take hold across different plant species at analogous moments.

106. Ibid., 54.

107. William Bateson, *Materials for the Study of Variation* (London: Macmillan, 1894); Hugo de Vries, *Die Mutationstheorie* (Leipzig: Veit, 1901–3).

108. Theodor Eimer, *Die Entstehung der Arten auf Grund von Vererben erworbener Eigenschaften nach den Gesetzen organischen Waschsens* (Jena: G. Fischer, 1888).

109. Edwin Drinker Cope, *The Primary Factors of Organic Evolution* (Chicago: Open Court, 1896), 123.

110. It is worth mentioning that this argument has been made against Bergson. See Robert Wesson, *Beyond Natural Selection* (Cambridge, MA: MIT Press, 1991), 189.

111. August Weismann, *Aufsätze über Vererbung und Verwandte Biologische Fragen* (Jena: G. Fischer, 1892).

112. Bergson, *Creative Evolution*, 89.

113. Ibid., 89.

114. Ibid., 91.

115. Ibid., 95–96.

116. Ibid., 252.

117. Ibid., 96.

118. Ibid., 230.

119. Jean-Marie Guyau, *A Sketch of a Morality Independent of Obligation or Sanction*, trans. Gertrude Kapteyn (London: Watts, 1898), 98; originally published as *Esquisse d'une morale sans obligation ni sanction* (Paris: Félix Alcan, 1885).

120. Ibid., 86.

121. Ibid., 76.

122. Ibid., 90.

123. See Keith Ansell-Pearson, "Morality and the Philosophy of Life in Guyau and Bergson," *Continental Philosophy Review* 47, no. 1 (2014): 57–85.

124. Bergson, *Creative Evolution*, 126.

125. Henri Bergson, "Life and Consciousness," in *Mind-Energy*, 18; originally published as, "La conscience et la vie," in *L'énergie spirituelle* (Paris: Félix Alcan, 1919), 1–31.

126. Léon Brunschvicg, *L'expérience humaine et la causalité physique* (Paris: Félix Alcan, 1922), 570.

127. Bergson, *Creative Evolution*, 250.

128. Ibid., 51.

129. Ibid., 181.

130. Ibid., 118, my emphasis.

131. "Bergson to Le Roy, Oct. 30, 1912," in *Annales bergsoniennes II* (Paris: Presses universitaires de France, 2002), 474–75.

132. Bergson, *Creative Evolution*, 16.

133. Ibid., 105.

134. Camille Riquier makes this point forcefully in "Vie et liberté," in *L'évolution créatrice de Bergson*, ed. Arnaud François (Paris: Librairie J. Vrin, 2010), 125–66.

135. Maurice Merleau-Ponty, *In Praise of Philosophy*, trans. John Wild and James Edie (Evanston, IL: Northwestern University Press, 1963), 914–15; originally published as *Éloge de la philosophie* (Paris: Gallimard, 1953), 21.

136. Bergson, *Creative Evolution*, 263.

Chapter Six

1. The notion of "two Frances" has been invoked to characterize the irreconcilable enmity between Catholics and Republicans after Jules Michelet's *Tableau de la France* (1833). More recently, see James F. McMillan, "Catholic Christianity in France from the Restoration to the Separation of Church and State, 1815–1905," in *The Cambridge History of Christianity*, vol. 8, ed. Sheridan Gilley and Brian Stanley (Cambridge: Cambridge University Press, 2006), 217–32.

2. In his November 18, 1893, encyclical, Pope Leo XIII acknowledged that evolutionary theory contradicted the church's teaching that the earth began six thousand years ago: "Providentissimus Deus," *Papal Encyclicals Online*, http://www.papal encyclicals.net/Leo13/l13provi.htm (accessed February 1, 2014).

3. Édouard Le Roy, "Discussion sur les éléments chrétiens de la conscience contemporaine," *Bulletin de la Société française de philosophie* 2, no. 3 (1902): 51.

4. George Tyrrell, quoted in Anthony M. Maher, *The Forgotten Jesuit of Catholic Modernism: George Tyrrell's Prophetic Theology* (Minneapolis: Fortress, 2018), 38.

5. Le Roy's contributions to French spiritualism are widely acknowledged. Blondel's, however, are contestable. My argument in this chapter hews to that of Michael Conway, who states, "Blondel drew fundamentally on principles developed and expounded by the spiritual positivists in their able criticism of both mechanistic materialism and an evolutionary idealism that was emerging from across the Rhine." Conway, "Maurice Blondel on the Structures of Science within a Positive Phenomenology," *Irish Theological Quarterly* 69 (2004): 380. To the contrary, Jean Leclercq argues, "One cannot truly affirm that he claims an affiliation with this 'wave,' even if spiritualism affirms that man is capable, by his intelligence and by his will, of freeing himself from natural necessity, thanks to a progressive conquest over himself." Leclercq, "La logique de la vie: Lectures du 'jeune' Maurice Blondel (1881–1893)," vol. 2 (PhD diss., Université Catholique de Louvain, 2002), 23.

6. Quoted in André Lalande, "Pragmatisme et pragmaticisme," *Revue philosophique* 61 (1906): 123.

7. Blondel refers to his commitment to pragmatism in "Lettre," *Revue du clergé français* 29 (1902): 652.

8. Édouard Le Roy, *Dogme et critique* (Paris: Bloud et Cie, 1907), 86.

9. Oliva Blanchette recounts Blondel's decision not to republish *L'action* in *Maurice Blondel: A Philosophical Life* (Grand Rapids, MI: Eerdmans, 2010), 285–88.

10. Bergson's close relation to James, as well as the coemergence of pragmatism in France and America, has been well documented. See David G. Schultenover, SJ, ed. *The Reception of Pragmatism and the Rise of Roman Catholic Modernism, 1890–1914* (Washington, DC: Catholic University of America Press, 2009); Stéphane Madelrieux, ed., *France Bergson et James, cents ans après* (Paris: Presses universitaires de France, 2011), especially Paola Marrati, "James, Bergson et un univers en devenir," 123–40; Larry McGrath, "Bergson Comes to America," *Journal of the History of Ideas* 74, no. 4 (2013): 599–620.

11. William James, *Pragmatism: A New Name for Some Old Ways of Thinking* (New York: Longmans, Green, 1907). In the preface, James cites Blondel and Le Roy as influences, p. viii. Bergson also wrote an introduction for the book's French translation: William James, *Le pragmatisme*, trans. E. Le Brun (Paris: Flammarion, 1911).

12. John Dewey, *The Quest for Certainty: A Study of the Relation of Knowledge and Action* (New York: Minton, Balch, 1929).

13. Biran's unfinished manuscript was titled "Les fondements de la morale et de la religion." His oeuvre has been categorized into three stages: sensationalist, stoic, and mystic or religious. See Gouhier, *Les conversions de Maine de Biran*; Georges Le Roy, *L'expérience de l'effort et de la grâce chez Maine de Biran* (Paris: Boivin, 1937).

14. A. de Lavalette-Monbru, ed., *Journal intime de Maine de Biran*, vol. 2 (Paris: Plon, 1927), 104.

15. Ethics lessons in the 1832 *programme* included "The principal attributes of God; divine Providence, and the plan of the universe," and "Religious ethics, or duties to God." Victor Cousin, *Défense de l'université*, 3rd ed. (Paris: Joubert, 1844), 362.

16. D. Greer, *The Incidence of the Terror during the French Revolution: A Statistical Interpretation* (Cambridge, MA: Harvard University Press, 1935), 161–63.

17. Joseph F. Byrnes, *Catholic and French Forever: Religious and National Identity in Modern France* (University Park: Pennsylvania State University Press, 2005).

18. New French journals included *La revue Thomiste* (1893); *La revue néo-scolastique de philosophie* (1894); the Jesuit periodical *Etudes* (1897); and *La revue de sciences philosophiques et théologiques* (1907). Already existing Catholic journals included *Annales de philosophie chrétienne* and *La revue de philosophie.*

19. Alfred Loisy, *The Gospel and the Church*, trans. Christopher Home (New York: Charles Scribner's Sons, 1904), 19; originally published as *L'évangile et l'église* (Paris: Bellevue, 1902). Following the church's initial condemnation of Loisy's historicist approach in 1902, he further defended his "little book" the next year in *Autour d'un petit livre* (Paris: Alphonse Picard, 1903).

20. Ibid., 87.

21. See John 2:13 and John 6:4.

22. Loisy, *Gospel and the Church*, 38.

23. See John Henry Newman, *An Essay on the Development of Christian Doctrine* (London: James Toovey, 1845).

24. See Marvin R. O'Connell, *Critics on Trial: An Introduction to the Catholic Modernist Crisis* (Washington, DC: Catholic University of America Press, 1994).

25. Pope Pius X, "Lamentabili Sane Exitu (3 July 1907)," *Papal Encyclicals Online*, http://www.papalencyclicals.net/Pius10/p10lamen.htm (accessed February 1, 2014).

26. Ibid.

27. Pope Pius X, "Pascendi Dominici Gregis (8 Sept. 1907)," *Papal Encyclicals Online*, http://www.papalencyclicals.net/Pius10/p10pasce.htm (accessed February 1, 2014).

28. Ibid.

29. Ibid.

30. Roger D. Haight, "The Unfolding of Modernism in France: Blondel, Laberthonnière, Le Roy," *Theological Studies* 35 (1974): 633.

31. Alfred Loisy, *My Duel with the Vatican*, trans. Richard Wilson Boynton (New York: E. P. Dutton, 1924), 319; originally published as *Choses passées* (Paris: Émile Nourry, 1913).

32. Pope Pius X, "Motu Proprio—Sacrorum Antistitum," *Papal Encyclicals Online*, https://www.papalencyclicals.net/pius10/p10moath.htm (accessed September 5, 2019).

33. Biographical details found in Blanchette, *Maurice Blondel*, 32–42.

34. See George H. Tavard, "Blondel's *Action* and the Problem of the University," in *Catholicism Contending with Modernity: Roman Catholic Modernism and Anti-modernism in Historical Context*, ed. Darrell Jodock (Cambridge: Cambridge University Press, 2000), 142–68.

35. See Joel Revill, "Émile Boutroux, Redefining Science and Faith in the Third Republic," *Modern Intellectual History* 6, no. 3 (2009): 485–512.

36. Later in his career, Boutroux addressed Catholicism in *Science et religion dans la philosophie contemporaine* (Paris: Ernest Flammarion, 1908).

37. Émile Boutroux, letter to Maurice Blondel, July 16, 1884, CIV.81, Archives de Maurice Blondel, Université Catholique de Louvain, Louvain-La-Neuve, Belgium.

38. Maurice Blondel, *Action (1893): Essay on a Critique of Life and a Science of Practice*, trans. Oliva Blanchette (Notre Dame, IN: University of Notre Dame Press, 1984), 3; originally published as *L'action: Essai d'une critique de la vie et d'une science de la pratique* (Paris: Félix Alcan, 1893).

39. Ibid., 5.

40. Le Roy, "Sur quelques objections adressées à la nouvelle philosophie," 316–17.

41. Blondel, *Action*, 65.

42. Maurice Blondel, *Léon Ollé-Laprune: L'achèvement et l'avenir de son œuvre* (Paris: Bloud et Gay, 1923), 69. The book was originally written as a eulogy in 1899.

43. Ibid.

44. Édouard Le Roy, "Science et philosophie (III)," *Revue de métaphysique et de morale* 7, no. 6 (1899): 715.

45. Maurice Blondel, "Une association inséparable: L'agrandissement des astres à l'horizon," *Revue philosophique* 26 (1888): 489–97. The article was published per the recommendation of Boutroux, who wrote to Blondel: "I find your work very interesting and ingenious . . . and I'll send it to Mr. Ribot, asking him—which I'm sure is unnecessary—to welcome it for his journal." Émile Boutroux, letter to Maurice Blondel, July 17, 1888, CIV.80, Archives de Maurice Blondel, Université Catholique de Louvain.

46. The argument was prescient for having articulated, in an inchoate manner, the relativity of light and gravitation that Albert Einstein later elaborated. In fact, a British astronomical expedition to Africa in 1919 verified Blondel's claim. Photographs of a solar eclipse showed that light is affected by the earth's gravitational pull and thus acts like any other mass. See Adam C. English, "'Science Cannot Stop with Science': Maurice Blondel and the Sciences," *Journal of the History of Ideas* 69, no. 2 (2008): 272.

47. Blondel, *Action*, 165.

48. Ibid.

49. Michael Conway, *The Science of Life: Maurice Blondel's Philosophy of Action and the Scientific Method* (New York: Peter Lang, 2000).

50. Émile Boutroux, "Rapport au doyen de la faculté des lettres de Paris" (July 27, 1892), in Maurice Blondel, *Lettres philosophiques* (Paris: Aubier, 1961), 21.

51. Blondel, *Action*, 66.

52. Ibid., 88.

53. English, "'Science Cannot Stop with Science,'" 289.

54. Blondel, *Action*, 92–93.

55. Ibid., 90.

56. Le Roy, "Science et philosophie (II)," 506.

57. Henri Poincaré, *Science et méthode* (Paris: Flammarion, 1908), 131.

58. Blondel, *Action*, 106.

59. Ibid.

60. Ibid.

61. Ibid., 123.

62. The examples are borrowed from John Milbank's reading of Blondel in *Theology and Social Theory: Beyond Secular Reason*, 2nd ed. (Oxford: Blackwell, 2006), 214.

63. Ibid., 141.

64. See William Portier, "Twentieth-Century Catholic Theology and the Triumph of Maurice Blondel," *Communio* 38, no. 1 (2011): 103–37.

65. Blondel, *Action*, 87.

66. Blondel's critique of Kant played a formidable role in the development of his dissertation. He wrote to Boutroux, "I contemplated defining it otherwise than in Kantianism, with its relations between speculation and practice, between knowledge and existence; above all, I contemplated showing the indestructible reality and radical insufficiency of the entire natural order, in order to discover the flaw in what I believe had been separated." Maurice Blondel, letter to Émile Boutroux, no date, CXII.28, Archives de Maurice Blondel, Université Catholique de Louvain.

67. Blondel, *Action*, 325.

68. Ibid.

69. Søren Kierkegaard, *Sickness unto Death: A Christian Psychological Exposition for Upbuilding and Awakening*, 1849, trans. Howard B. Hong and Edna H. Hong (Princeton, NJ: Princeton University Press, 1980), 21.

70. Blondel, *Action*, 321.

71. Ibid., 328.

72. Ibid., 345–46.

73. Abbé Charles Denis, "Les idées et les hommes: I. Nouvelles tendances de l'apologétique philosophique; MM. Ollé-Laprune, George Fonsegrive, Maurice Blondel," *Annales de philosophie chrétienne* 32 (1895): 656. Blondel later took over the journal in 1905 following Denis's death.

74. Maurice Blondel, *The Letter on Apologetics and History and Dogma*, trans. Alexander Dru and Illtyd Trethowan (Grand Rapids, MI: Eerdmans, 1994), 152; originally published as "Lettre sur les exigences de la pensée contemporaine en matière d'apologétique et sur la méthode philosophique dans l'étude du problème religieux," *Annales de philosophie chrétienne* 131 (1896): 337–47, 467–82, 599–616; 132 (1896): 225–67, 337–50.

75. Ibid., 181.

76. Ibid., 159–60.

77. Ibid., 134.

78. Jean Leclercq, "Du Dieu qui vient à l'action: Le dénouement blondélien," *Revue philosophique de Louvain* 99, no. 3 (2001): 422–52.

79. Blondel, *Action*, 318.

80. See Ralph Gibson, *A Social History of French Catholicism, 1789–1914* (London: Routledge, 1989).

81. Pope Pius X, "Lamentabili Sane Exitu," Proposition 26.

82. Ibid.

83. See Pope Pius IX, "Ineffabilis Deus (8 Dec. 1854)," *Papal Encyclicals Online*, http://www.papalencyclicals.net/Pius09/p9ineff.htm (accessed February 1, 2014).

84. See Pope Pius XII, "Munificentissimus Deus (1 Nov. 1950)," *Papal Encyclicals Online*, http://www.papalencyclicals.net/Pius12/P12munif.htm (accessed February 1, 2014).

85. Loisy, *Gospel and the Church*, 218.

86. Ibid., 224.

87. The articles were originally conceived in 1897 as part of a larger project, *The Christian Spirit* (*L'esprit chrétien*), which Blondel would not publish until 1944.

88. Père René Marlé, *Au cœur de la crise moderniste* (Paris: Aubier, 1960), 201.

89. Maurice Blondel, "History and Dogma," in *Letter on Apologetics and History and Dogma*, 269; originally published as "Histoire et dogme: Les lacunes philosophiques de l'exégèse moderne," *La quinzaine* 56 (1904): 145–67, 349–73, 433–58.

90. Blondel, "Lettre sur les exigences de la pensée contemporaine," 78.

91. Blondel, "History and Dogma," 275.

92. Loisy, *Gospel and the Church*, 87.

93. Blondel, "History and Dogma," 247.

94. Ibid., 267.

95. Émile Boutroux, letter to Maurice Blondel, August 20, 1905, CIV.98, Archives de Maurice Blondel, Université Catholique de Louvain.

96. Blondel, "History and Dogma," 281.

97. Édouard Le Roy, *What Is a Dogma?*, trans. Lydia G. Robinson (Chicago: Open Court, 1918), 83; originally published as "Qu'est-ce qu'un dogme?," *La quinzaine* 63 (1905): 495–526; Le Roy outlined his contributions to the problem of dogma as early as 1902 in "Discussion sur les éléments chrétiens de la conscience contemporaine," 62.

98. Le Roy, *What Is a Dogma?*, 28.

99. Alfred Loisy, letter to Édouard Le Roy, June 6, 1905, 1NA1 Art. 16, Fonds Édouard Le Roy, Institut Catholique de Paris.

100. Le Roy, *What Is a Dogma?*, 71.

101. Abbé Wehrlé, letter to Édouard Le Roy, July 5, 1905, 1NA1 Art. 16, Fonds Édouard Le Roy, Institut Catholique de Paris.

102. G. Letourneux, letter to Édouard Le Roy, May 12, 1905, ibid.

103. See Lucien Laberthonnière, "Le problème religieux," *Annales de philosophie chrétienne* 132 (1897): 497–511, 615–32.

104. Maurice Blondel, letter to Lucien Laberthonnière, November 10, 1905, XXIX.3, Archives de Maurice Blondel, Université Catholique de Louvain.

105. Le Roy, *What Is a Dogma?*, 70.

106. Ibid., 88.

107. Maurice Blondel, letter to Édouard Le Roy, November 16, 1905, in Blondel, *Lettres philosophiques*, 258.

108. Ibid.

109. Le Roy, *What Is a Dogma?*, 79.

110. Édouard Le Roy, "Scolastique et philosophie moderne," *Demain*, August 15 1906.

111. Henri Bergson, letter to Édouard Le Roy, May 31, 1905, 1NA1 Art. 16, Fonds Édouard Le Roy, Institut Catholique de Paris.

112. Le Roy, "Discussion sur les éléments chrétiens de la conscience contemporaine," 54.

113. Ibid., 52.

114. Eugène Lenoble, "Revue de L'évolution créatrice," *Revue du clergé français* 53 (1908): 180–208.

115. Réginald Garrigou-Lagrange, *Le sens commun, la philosophie de l'être et les formules dogmatiques* (Paris: Beauchesne, 1909).

116. Lucien Laberthonnière, *Essais de philosophie religieuse* (Paris: Lethielleux, 1903); Laberthonnière, *Le réalisme chrétien et l'idéalisme grec* (Paris: Lethielleux, 1904).

117. Maurice Blondel and Lucien Laberthonnière, *Correspondance philosophique* (Paris: Éditions du Seuil, 1961), 192.

118. See Blanchette, *Maurice Blondel*, 257.

119. Notably, Testis, "La semaine sociale de Bordeaux et le monophorisme: Controverses sur les méthodes et les doctrines," *Annales de philosophie chrétienne* 159 (1909): 5–22, 162–84, 245–78; (1910): 372–92, 449–72, 561–92; 160 (1911): 127–62.

120. Le Roy acknowledged his debt to Blondel in *Dogme et critique*, 70. Blondel subsequently disavowed any likeness between his thought and Le Roy's in "L'apologétique et la philosophie de M. Blondel," *Revue du clergé français* 50 (1907): 546.

121. Charles Péguy, *Note sur M. Bergson et la philosophie bergsonienne: Note conjointe sur M. Descartes et la philosophie cartésienne* (Paris: Éditions Gallimard, 1933).

122. Quoted in Blaise Romeyer, "Caractéristiques religieuses du spiritualisme de Bergson," in *Bergson et le bergsonisme*, ed. Édouard Le Roy (Paris: Beauchesne, 1947), 29.

123. Quoted in Gabriel Flynn, "Introduction," in *Ressourcement: A Movement for Renewal in Twentieth-Century Catholic Theology*, ed. Gabriel Flynn and Paul Murray (Oxford: Oxford University Press, 2012), 4.

124. Pope John Paul II's letter to Archbishop Bernard Panafieu of Aix en Provence, translated by Mark Sebanc in "Notes and Comments, on the Centenary of Blondel's *L'action*," *Communio* 20 (Winter 1993): 721–23, quote p. 722.

125. See James Chapel, *Catholic Modern: The Challenge of Totalitarianism and the Remaking of the Church* (Cambridge, MA: Harvard University Press, 2018).

Epilogue

1. Ravaisson, *Rapport sur la philosophie en France au XIX*^e^ *siècle*, 258.

2. Émile Boutroux, "La philosophie en France depuis 1867," *Revue de métaphysique et de morale* 16, no. 6 (1908): 685.

3. Ibid.

4. Ibid., 714.

5. "Human Brain Project," 2017, http://www.humanbrainproject.eu.

6. See Christophe Prochasson and Anne Rasmussen, *Au nom de la patrie: Les intellectuelles et la première guerre mondiale (1910–1919)* (Paris: Editions de la Découverte, 1996).

7. Émile Boutroux, "L'Allemagne et la guerre Lettre de M. Émile Boutroux," *Revue des deux mondes* 23 (October 15, 1914): 395–401, quote p. 395.

8. Henri Bergson, "La signification de la guerre," in *La signification de la guerre* (Paris: Bloud et Gay, 1915), 15; cited in Martha Hanna, *Mobilization of the Intellect: French Scholars and Writers during the Great War* (Cambridge, MA: Harvard University Press, 1996), 92.

9. Georges Friedmann, "Ils ont perdu la partie éternelle d'eux-mêmes," *Esprit*, no. 1 (1926): 169.

10. Georges Politzer, *La fin d'une parade philosophique: Le bergsonisme* (Paris: Les Revues, 1929), 31.

11. Paul Nizan, *Les chiens de garde* (Paris: Rieder, 1932).

12. Quoted in Robert C. Grogin, "Rationalists and Anti-rationalists in Pre–World War I France: The Bergson-Benda Affair," *Historical Reflections/Réflexions historiques* 5, no. 2 (1978): 225.

13. Michel Foucault, "La psychologie de 1850 à 1950," *Tableau de la philosophie contemporaine*, ed. Alfred Weber and Denis Huisman (Paris: Éditions Fishbacher, 1957), 591–604.

14. Louis Althusser, "The Philosophical Conjuncture and Marxist Theoretical Research" in *The Humanist Controversy and Other Writings (1966–67)*, trans. G. M. Goshgarian (New York: Verso, 2003), 5.

15. Ibid.

16. Bergson's lecture was subsequently published as *Durée et simultanéité: A propos de la théorie d'Einstein* (Paris: Félix Alcan, 1922). For historical treatments of the debate, see Élie During, *Bergson et Einstein: La querelle du temps* (Paris: Presses universitaires de France, 2020); Jimena Canales, *The Physicist and the Philosopher: Einstein, Bergson, and the Debate That Changed Our Understanding of Time* (Princeton, NJ: Princeton University Press, 2015).

17. For an incisive account of Bergson's reception in the twentieth century, especially among psychiatrists, see Giuseppe Bianco, *Après Bergson, portrait de groupe avec philosophe* (Paris: Presses universitaires de France, 2015).

18. Charles Blondel, *La conscience morbide: Essai de psychopathologie générale* (Paris: Félix Alcan, 1914).

19. Georges Dwelshauvers, *L'inconscient* (Paris: Flammarion, 1916).

20. Eugène Minkowski, *Le temps vécu: Étude phénoménologique et psychopathologique* (Paris: L. d'Artrey, 1933).

21. See Noemi Pizarroso López, "De la historia de la filosofía a la psicología del misticismo: Los primeros trabajos de Henri Delacroix," *Revista de historia de la psicología* 34, no. 1 (2013): 81–109; Frédéric Fruteau de Laclos, *La psychologie des philosophes: De Bergson à Vernant* (Paris: Presses universitaires de France, 2012).

22. See Henri Ellenberger, *The Discovery of the Unconscious: The History and Evolution of Dynamic Psychiatry* (New York: Basic Books, 1970).

23. See Pierre Janet, *Les médications psychologiques* (Paris: Félix Alcan, 1919).

24. See Jacqueline Carroy and Régine Plas, "How Pierre Janet Used Pathological Psychology to Save the Philosophical Self," *Journal of the History of the Behavioral Sciences* 36, no. 3 (2000): 231–40.

25. See Pierre Janet, *La force et la faiblesse psychologique* (Paris: Maloine, 1930).

26. Px Zu, X. Zhang, S. Heany, A. Yoon, A. M. Michelson, and R. L. Maas, "Regulation of Pax6 Expression Is Conserved between Mice and Flies," *Development* 126, no. 2 (1999): 383–95.

27. Bergson, *Creative Evolution*, 89.

28. See Robert Kozma and Walter Freeman, *Cognitive Phase Transitions in the Cerebral Cortex—Enhancing the Neuron Doctrine by Modeling Neural Fields* (Heidelberg: Springer, 2015).

29. Arturo Alvarez-Buylla, Marga Theelen, and Fernando Nottebohm, "Birth of

Projection Neurons in the Higher Vocal Center of the Canary Forebrain before, during, and after Song Learning," *Proceedings of the National Academy of Sciences of the United States of America* 85, no. 22 (1988): 8722–26; Steve Goldman and Fernando Nottebohm, "Neuronal Production, Migration, and Differentiation in a Vocal Control Nucleus of the Adult Female Canary Brain," *Proceedings of the National Academy of Sciences of the United States of America* 80, no. 8 (1983): 2390–94.

30. Peter S. Eriksson, Ekaterina Perfilieva, Thomas Björk-Eriksson, et al., "Neurogenesis in the Adult Human Hippocampus," *Nature Medicine* 4, no. 11 (1998): 1313–17.

31. Jean-Pierre Changeux, *The Physiology of Truth: Neuroscience and Human Knowledge*, trans. M. B. DeBevoise (Cambridge, MA: Belknap Press of Harvard University Press, 2002), 32; originally published as *L'homme de vérité* (Paris: Odile Jacob, 2002). Other neuroscientists have affirmed the relevance of Bergson's thought; see Phillippe Gallois and Gérald Forzy, eds., *Bergson et les neurosciences* (Le Plessis-Robinson: Institut Synthélabo pour le progrès de la connaissance, 1997); Andrew C. Papanicolaou and Pete A. Y. Gunter, eds., *Bergson and Modern Thought: Towards a Unified Science* (Chur, Switzerland: Harwood, 1987).

32. Joseph LeDoux, *The Deep History of Ourselves: The Four-Billion-Year Story of How We Got Conscious Brains* (New York: Viking, 2019), 9. LeDoux's claim echoes that which he originally made in *Synaptic Self: How Our Brains Become Who We Are* (London: Macmillan, 2003), 324.

33. For incisive histories of phenomenology's initial reception in France, see Ethan Kleinberg, *Generation Existential: Heidegger's Philosophy in France, 1927–1961* (Ithaca, NY: Cornell University Press, 2005), especially ch. 2; Samuel Moyn, *Origins of the Other: Emmanuel Levinas between Revelation and Ethics* (Ithaca, NY: Cornell University Press, 2005).

34. Emmanuel Levinas, *La théorie de l'intuition dans la phénoménologie de Husserl* (Paris: Vrin, 1930).

35. Maurice Merleau-Ponty, *L'union de l'âme et du corps chez Malebranche, Biran, et Bergson*, ed. Jean Deprun (Paris: Vrin, 1968).

36. Gabriel Marcel, *Le monde cassé* (Paris: Desclée de Brouwer, 1933).

37. Jean Wahl, *Vers le concrète: Études d'histoire de la philosophie contemporaine* (Paris: Vrin, 1932).

38. See John Hellman, *Emmanuel Mounier and the New Catholic Left, 1930–1950* (Toronto: University of Toronto Press, 1981).

39. From "Conférence de Madrid: La Personnalité" (1916) in Robinet, *Mélanges*, 1215–35.

40. See Samuel Moyn, *Christian Human Rights* (Philadelphia: University of Pennsylvania Press, 2015).

41. See Roy Wood Sellars, "The Spiritualism of Lavelle and le Senne," *Philosophy and Phenomenological Research* 11, no, 3 (1951): 386–93.

42. René Le Senne, *Obstacle et valeur: La description de conscience* (Paris: Fernand Aubier, 1934); Le Senne, *Introduction à la philosophie* (Paris: Félix Alcan, 1925).

43. Louis Lavelle, "De la relation de l'esprit et du monde," in *Actas del primer congreso nacional de pilosofía* (Mendoza, Argentina: Instituto de Filosofía, Universidad Nacional de Cuyo, 1949), 825.

44. Ibid.

45. Gilles Deleuze, "Afterword," in *Bergsonism*, 1988, trans. Hugh Tomlinson and

Barbara Habberjam (Brooklyn, NY: Zone Books, 2006); originally published as *Le bergsonisme* (Paris: Presses universitaires de France, 1966).

46. See Rebecca Hill, *The Interval: Relation and Becoming in Irigaray, Aristotle, and Bergson* (New York: Fordham University Press, 2014).

47. See Todd Cronan, *Against Affective Formalism: Matisse, Bergson, Modernism* (Minneapolis: University of Minnesota Press, 2014).

48. See Donna V. Jones, *The Racial Discourses of Life Philosophy: Négritude, Vitalism, and Modernity* (New York: Columbia University Press, 2010).

49. See Adriana Alfaro, "The Belief in Intuition: Personality and Authority in Henri Bergson and Max Scheler" (PhD diss., Harvard University, 2017); Alexander Lefebvre and Melanie White, eds., *Bergson, Politics, and Religion* (Durham, NC: Duke University Press, 2012).

50. See Alexander Lefebvre, *Human Rights and the Care of the Self* (Durham, NC: Duke University Press, 2018), ch. 5; *Human Rights as a Way of Life: On Bergson's Political Philosophy* (Stanford, CA: Stanford University Press, 2013); Nadia Yala Kisukidi, *Bergson ou l'humanité créatrice* (Paris: CNRS, 2013).

Index

Note: An *f* following a page number indicates a figure or figures; *n* indicates a note; *t* indicates a table.